INTELLECTUAL PROPERTY RIGHTS IN ANIMAL BREEDING AND GENETICS

Intellectual Property Rights in Animal Breeding and Genetics

Edited by

Max Rothschild
Department of Animal Science
Iowa State University, Ames, Iowa, USA

and

Scott Newman
PIC USA
Franklin, Kentucky, USA

CABI *Publishing*

CABI *Publishing* is a division of CAB *International*

CABI Publishing
CAB International
Wallingford
Oxon OX10 8DE
UK

Tel: +44 (0)1491 832111
Fax: +44 (0)1491 833508
E-mail: cabi@cabi.org
Web site: www.cabi.publishing.org

CABI Publishing
10 E 40th Street
Suite 3203
New York, NY 10016
USA

Tel: +1 212 481 7018
Fax: +1 212 686 7993
E-mail: cabi-nao@cabi.org

A catalogue record for this book is available from the British Library, London, UK.

Library of Congress Cataloging-in-Publication Data
Intellectual property rights in animal breeding and genetics / edited by Max Rothschild, Scott Newman.
p. cm.
Includes bibliographical references.
ISBN 0-85199-641-8 (alk. paper)
1. Livestock--Germplasm resources--Patents. 2. Intellectual property. I. Rothschild, Max Frederick, 1952- II. Newman, Scott.
SF105.3 .I67 2002
636.08′21--dc21 2002006037

ISBN 0 85199 641 8

Typeset in Baskerville by Columns Design Ltd, Reading.
Printed and bound in the UK by Biddles Ltd, Guildford and King's Lynn.

Contents

Contributors

Lila Akrad is an Associate at the law firm of McKee, Voorhees & Sease. She received a J.D. from Vermont Law School, South Royalton, Vermont, and attended the College of Veterinary Medicine, Iowa State University, Ames. She also received a BS in organismal biology from the University of Kansas, Lawrence. She was a Research Assistant, Alaskan Expedition, for the California Academy of Sciences, and was an Administrative Assistant at the Howard Hughes Medical Institute, Iowa City, Iowa. Current address: McKee, Voorhees & Sease, 801 Grand Avenue, Suite 3200, Des Moines, IA 50309-2721, USA. akrad@ipmvs.com

Bill Barendse is Principal Research Scientist with CSIRO Livestock Industries. He studies genes that affect normal variation; the everyday shape, size and compositional differences between individual animals. He began his career studying evolutionary genetics while at the University of Western Australia. His research includes studying a human congenital disorder while at Macquarie University and now at CSIRO studying the genes contributing to normal variation for meat quality in cattle. He has helped to construct some of the tools for this effort which has led to the construction of a genetic map of cattle (http://www.cgd.csiro.au/). DNA tests based on this research are now available. Current address: Molecular Animal Genetics Centre, CSIRO Livestock Industries, Level 3 Gehrmann Laboratories, Research Road, University of Queensland, St Lucia, QLD 4072, Australia. Bill.Barendse@li.csiro.au

Andrew Blattman joined Spruson & Ferguson in 1995 and is a member of the Chemical Patent Group, specializing in biotechnology patents.

He holds a Bachelor of Science degree in agriculture, majoring in genetics and molecular biology. Later, while working within the CSIRO Division of Animal Health, he completed a PhD in the molecular genetics of parasite resistance. Immediately prior to joining the firm, he undertook at the University of Wisconsin-Madison an 18-month post-doctoral fellowship studying the isolation of quantitative trait loci and marker-assisted selection. He has completed courses in nucleic acid amplification, cloning, sequencing and quantitative genetics. Andrew is the author of a number of research articles in international journals, scientific conference papers, monographs and book chapters. He became a Principal of the firm in 1999. He also lectures in Patent Systems at the University of Technology Sydney, and is a co-author of the Biotechnology IP Management Manual commissioned by the Australian government agency, Biotechnology Australia. Current address: Spruson & Ferguson, Level 35/31 Market Street, Sydney, NSW 2000, Australia. andrew.blattman@sprusons.com.au

Colin Bodkin joined Spruson & Ferguson in 1991 after an accomplished career in industry. His past employment includes research and teaching appointments at the Universities of New South Wales and Adelaide, in addition to positions as senior research officer with Carlton & United Breweries, chief chemist with CSR Chemicals, technical manager with ICI Biospecialities and technical manager with Marblo Ltd. His doctorate was taken in organic chemistry. Colin provides patent advice to clients in the areas of analytical and organic chemistry, industrial chemistry and biotechnology. He lectures in patent systems at the University of Technology Sydney and has been Principal of the firm since 1996. He has also contributed to Halsbury's Laws of Australia, and is a co-author of the Biotechnology IP Management Manual commissioned by the Australian government agency, Biotechnology Australia. Current address: Spruson & Ferguson, Level 35/31 Market Street, Sydney, NSW 2000, Australia. colin.bodkin@sprusons.com.au

Tonya Brady is an Associate at the law firm of McKee, Voorhees & Sease. She received a J.D. from the University of Dayton Law School, Dayton, Ohio, and a B.S. in molecular genetics from the Ohio State University, Columbus. She was employed at the Ohio State University Medical School Department of Radiology as a research assistant and as a laboratory assistant in the Department of Molecular Genetics. She has numerous publications in the field of molecular carcinogenesis. Current address: McKee, Voorhees & Sease, 801 Grand Avenue, Suite 3200, Des Moines, IA 50309-2721, USA. brady@ipmvs.com

José Cibelli graduated from the School of Veterinary Medicine at University of La Plata in Argentina in 1989. In 1998, he finished his

PhD in reproductive physiology. Dr Cibelli is one of the pioneers in the area of cloning with transgenic somatic cells for the production of animals and embryonic stem cells. His work was focused in the production of transgenic cattle and indeed, in January 1998, Dr Cibelli's efforts led to the generation of the world's first transgenic calves by cloning. This was followed by high-profile publications in *Science, Nature Biotechnology* and *Nature Medicine.* Dr Cibelli is the editor of the book *Principles of Cloning*, to be published by Academic Press. Currently, Dr Cibelli is Vice-President for Research at Advanced Cell Technology in Worcester, Massachusetts. His major area of research is cloning and stem cells for their applications in agriculture and medicine. Current address: Advanced Cell Technology, One Innovation Drive, Biotech Three, Worcester, MA 01605, USA. jcibelli@advancedcell.com

Margaret Dunbar is a registered patent attorney currently working as a consultant. She has spent the majority of the last several years working as a consultant supporting the Animal Agriculture Groups of Monsanto Company, with much of that support to their genomics groups and projects. She has a B.S. in animal science from the University of California at Davis, and an M.S. in animal science and J.D., both from the University of Idaho. Prior to embarking on her career as a Patent Attorney she worked as a research assistant in large animal reproduction at the University of California at Davis. Current address: Cypress Bioscience, Inc., 4350 Executive Drive, San Diego, CA 92130, USA. mdunbar@cypressbio.com

Professor Donald Evans is Director of the Bioethics Centre at the University of Otago. He graduated summa cum laude in philosophy in the University of Wales where he also completed his PhD in moral philosophy. He taught at the University for 29 years. He founded the Centre for Philosophy and Health Care and developed the first European Masters degree programme in medical ethics. He is a member of the governing body of the Institute of Medical Ethics and has published seven authored and edited books and more than 50 learned papers and chapters in books. He was elected a member of the Academy of Humanitarian Research, Moscow, in 1996. He has conducted research projects for the UK Department of Health on the ethical review of clinical research, and coordinated two European Commission concerted actions. He has completed two research reports on health care systems in the Third World, one for the European Commission and the other for the World Health Organization, and a further research report on resource allocation for the New Zealand Ministry of Health. He is currently leading a WHO project on the development of ethical review of research involving human subjects in South East Asia, South America and Africa. He is a member of the Independent Biotechnology Advisory

Committee of the New Zealand Ministry of Research, Science and Technology. He served for many years as a magistrate in the UK. He has chaired and served on numerous ethics committees including those of the British Medical Association, the Royal College of Nursing, the Association of Clinical Research Contractors and the Health Research Council of New Zealand. Current address: University of Otago Medical School, Dunedin Hospital, Dunedin, New Zealand. Donald.Evans@stonebow.otago.ac.nz

Thomas A. Fretz is Dean of the College of Agriculture and Natural Resources, and director of both the Maryland Agricultural Experiment Station and Maryland Cooperative Extension, at the University of Maryland, College Park. Prior to serving as Dean at the University of Maryland, Dr Fretz served as the Associate Dean of the College of Agriculture and Associate Director of the Iowa Agriculture and Home Economics Experiment Station at Iowa State University where he was responsible for the direction of statewide agriculture research programmes. Previously he served on the faculty at the University of Georgia in the Georgia Experiment Station, The Ohio State University, Kansas State University and Virginia Polytechnic Institute, where he served as chair of the Department of Horticulture from 1981 to 1989. Current address: University of Maryland, Office of the Dean, College of Agriculture, 1104 Symons Hall, College Park, MD 20742, USA. tf43@umail.umd.edu

Bruce Golden is the founder and Coordinator of the Colorado State University Center for Genetic Evaluation of Livestock and a Professor of Animal Breeding and Genetics at CSU. He is also a founder, chairman of the board and Chief Scientific Officer of Optibrand Ltd, LLC, a company commercializing products in secure animal source verification. Dr Golden is the founder of the animal geneticist's discussion group, a list server with over 1100 animal geneticists from around the world. Dr Golden's research includes the development of software methods for national beef genetic evaluation and animal identification methods. His laboratory has developed new procedures for predicting genetic merit for beef female reproduction, finish endpoints, maintenance energy requirements and variance components estimation. He holds patents in animal identification that include the use of GPS information in animal identification. He received his PhD in animal breeding/genetics from Colorado State University. Current address: Department of Animal Sciences, Colorado State University, Fort Collins, CO 80523, USA. Bruce.Golden@ColoState.edu

Daniel Kevles is the Stanley Woodward Professor of History at Yale University. He has written extensively about issues in science and society past and present. His works include *In the Name of Eugenics: Genetics and the Uses of Human Heredity* (1985, 1995); *The Baltimore*

Case: a Trial of Politics, Science, and Character (1998); and, co-edited with Leroy Hood, *The Code of Codes: Scientific and Social Issues in the Human Genome Project*. He is currently writing a book on the history of intellectual property in living organisms. Current address: Department of History, Yale University, PO Box 208324, New Haven, CT 06520–8324, USA. daniel.kevles@yale.edu

Brian Kinghorn is Professor of Animal Breeding Technologies in the Department of Animal Science at the University of New England. He has experience across a variety of animal industries. His research emphasis is on design of breeding programmes, including optimal use of novel molecular and reproductive techniques. He has delivered a number of postgraduate courses in other countries. He has also supervised over 50 PhD and Masters research students. A recent industry development of his is the establishment of method, software and a service to industry for the implementation of breeding programmes, giving proper integration of the technical, logistical and cost issues facing animal breeders. Current address: Department of Animal Science, University of New England, Armidale, NSW 2351, Australia. bkinghor@metz.une.edu.au

Corinne Langinier is Adjunct Assistant Professor of economics at Iowa State University. She received her PhD from the University of Toulouse (France) in Economics in 1997. Her research area is the economics of patents and intellectual property rights. Her research focuses on the strategic behaviour of innovators, licensing and patent policy issues. Current address: Department of Economics, Iowa State University, 381 Heady Hall, Ames, IA 50011, USA. langinier@econ.iastate.edu

William H. Lesser is a Professor in the Department of Applied Economics and Management at Cornell University and has been there since receiving his PhD in agricultural economics with a specialization in marketing from the University of Wisconsin in 1978. Early on at Cornell he was an innovator in the application of PCs to food distribution, writing some of the earliest specialized software. Much of the time has, however, been focused on the farm and consumer level effects of biotechnology on agriculture. A particular specialization is the ramifications of patents and Plant Breeders' Rights. In a related area, he has examined ownership of and access to genetic resources. His work has involved advising the governments of Brazil, Bangladesh, Switzerland and Indonesia, among others. He has written three books and numerous articles and chapters on the subject of agricultural biotechnology. Current address: Department of Applied Economics and Management, Cornell University, 405 Warren Hall, Ithaca, NY 14853-7801, USA. whl1@cornell.edu

David R. MacKenzie is Executive Director of the Northeastern Regional Association of State Agricultural Experiment Station Directors

(NERA), headquartered at the University of Maryland in College Park. Prior to joining NERA in July 1995, Dr MacKenzie was National Programme Leader for the US Department of Agriculture's (USDA) Cooperative State Research, Education, and Extension Service (CSREES), located in Washington, DC. In this capacity, he managed the agency's biotechnology research programmes, giving emphasis to the safe application of the new tools of molecular biology to crop and livestock improvement. Dr MacKenzie is Chair of the Board of Trustees of the International Potato Center (CIP) located in Lima, Peru. He is also a long-time consultant to the World Bank in agriculture and natural resources management. Current address: University of Maryland, Agricultural Experiment Station, 120 Hartwick Building, College Park, MD 20742-3131, USA. dmackenz@umd.edu

John McCann practises law in the area of chemical, electrochemical, biochemical and physical chemistry patents. John has extensive experience in this field and holds a PhD for research in the fields of chemistry, electrochemistry, physics, solar cells, battery technology and lasers. He has also completed a course in quantum chemistry and solid-state physics at the University of Uppsala, Sweden. Prior to joining Spruson & Ferguson in 1982, John worked as a research and development chemist at Abbott Laboratories, before undertaking 3 years' postdoctoral experience in electrochemistry, solid state physics and solar cells as a research associate with the University of New South Wales' Physics Department. He is a member of the New York Academy of Sciences and is author and co-author of several articles in scientific journals, as well as chapters of scientific books. He is a member of the Royal Australian Chemical Institute and the American Association for the Advancement of Science. John became a principal of Spruson & Ferguson in 1988. He continues to be an examiner in patent law in the examinations for registration as a patent attorney, and lectures in professional conduct at the University of Technology, Sydney. He has contributed to Halsbury's Laws of Australia, and is a co-author of the Biotechnology IP Management Manual commissioned by the Australian government agency, Biotechnology Australia. Current address: Spruson & Ferguson, Level 35/31 Market Street, Sydney, NSW 2000, Australia. john.mccann@sprusons.com.au

Susan Meszaros is a Research Fellow in the School of Rural Science and Agriculture at the University of New England, from where she received her PhD. Her interest is in the area of optimization of breeding programmes and software development. She leads the development of a software system for the optimization of breeding programmes through mate selection together with Professor Brian Kinghorn. Current address: Department of Animal Science,

University of New England, Armidale, NSW 2351, Australia. smeszaro@metz.une.edu.au

GianCarlo Moschini is Professor of Economics and holds the Pioneer Hi-bred International Chair in Science and Technology Policy at Iowa State University. He holds a PhD degree from the University of Guelph and is a former editor of the *American Journal of Agricultural Economics*. His current research interests include the economics of R&D, biotechnology innovations and intellectual property rights. Current address: Department of Economics, Iowa State University, 381 Heady Hall, Ames, IA 50011, USA. Moschini@iastate.edu

Martha Mutschler is a Professor in the Department of Plant Breeding, College of Agriculture and Life Sciences, Cornell University. She directs a research programme in tomato and onion breeding and genetics. This work includes plant genetics/breeding projects concerning the genetic control of novel traits derived from wild species, the genetic control/physiological mechanisms underlying these traits and their use in vegetable improvement. This work has produced several US patents, as well as breeding lines with novel forms of disease resistance, insect resistance and modified production traits such as extended shelf-life and early maturity. Dr Mutschler has served on board of directors for the Cornell Research Foundation (the patent and licensing unit for Cornell University) for over a decade. Current interests include changing strategies for protecting plant intellectual property. She recently developed a computer-assisted instruction module for training undergraduate and graduate students in intellectual property issues within a university. Current address: Cornell University, Department of Plant Breeding, 303 Bradfield Hall, Ithaca, NY 14853, USA. mam13@cornell.edu

Jovanka Naumoska is a part-time student (science/law) employee in the firm of Spruson & Ferguson.

Heidi S. Nebel is a Partner at the law firm of McKee, Voorhees & Sease, where she practices exclusively in the area of biotechnological intellectual property. She holds a BS in molecular biology from the University of Iowa and a J.D. with honours from Drake University Law School. She is President Elect of the Iowa Intellectual Property Law Association and is a member of the Iowa Biotechnology Association, American Intellectual Property Law Association and American Bar Association. She is admitted to practice before the United States Patent and Trademark office, the Iowa Supreme Court, the Northern and Southern Districts of Iowa, the Federal Circuit Court of Appeals and the United States Supreme Court. Her practice includes heavy emphasis in all aspects of agricultural biotechnology. She prosecuted US Patent 6,136,320 (edible vaccines), which received nationwide acclaim in MIT's technology review, May 2000, as one of the five most significant patents of 2000; has published articles in *Iowa Commerce* magazine and the *Federal Circuit Bar Journal* and has represented both corporate and university inventors in all areas of biotechnology before the patent office and courts. Current address: McKee, Voorhees & Sease, 801 Grand Avenue, Suite 3200, Des Moines, IA 50309-2721, USA. nebel@ipmvs.com

Scott Newman is a Technical Services Geneticist with PIC USA. He received his PhD in animal breeding from Purdue University. He has held research and leadership positions at the Waite Agricultural Research Institute of the University of Adelaide, Agresearch New Zealand, USDA Agricultural Research Service, and in various divisions of the Commonwealth Scientific and Industrial Research Organization (CSIRO). Prior to joining PIC he was Genetics Program Leader in the Cooperative Research Centre for Cattle and Beef Quality. Current address: PIC USA, PO Box 348, Franklin, KY 42135, USA. scott.newman@pic.com

Graham Plastow is Chief Technology Officer for Sygen International Group. He trained as a molecular biologist, initially working on bacteria and plants. He initiated PIC's genomics programme in the late 1980s while working at the corporate laboratory of the then parent company Dalgety plc. Current address: Sygen International, Fyfield Wick, Abingdon, Oxfordshire OX13 5NA, UK. Graham.Plastow@pic.com

Max Rothschild is C.F. Curtiss Distinguished Professor of Agriculture at Iowa State University. He is leader of the Animal Breeding and Genetics group at ISU and USDA Pig Genome Coordinator. He has received several awards including the American Society of Animal Science Animal Breeding Award, USDA Honor award, an R&D100 award and is a fellow of American Association for the Advancement of Science. He holds several US and foreign patents. Current address: Department of Animal Science, Iowa State University, 2255 Kildee Hall, Ames, IA 50011, USA. mfrothsc@iastate.edu

Lawrence Schaeffer is Professor of Animal Breeding and Genetics at the University of Guelph, where he has been a member of the Animal and Poultry Science faculty since 1975. His research expertise focuses on dairy cattle genetic evaluation systems. He was awarded the Jay L. Lush Award in Animal Breeding and Genetics by the American Dairy Science Association in 1988. Current address: Centre for Genetic Improvement of Livestock, Department of Animal and Poultry Science, University of Guelph, Guelph, Ontario N1G 2W1, Canada. lrs@sherlock.aps.uoguelph.ca

Robin L. Teskin is a Partner in the McLean, Virginia office of Pillsbury Winthrop LLP specializing in intellectual property. She has been in private practice for 11 years. Her clients include numerous established and start-up biotechnology companies and universities including Advanced Cell Technology and IDEC Pharmaceuticals. Prior to private practice, Robin spent 8 years at the United States Patent and Trademark Office (USPTO) as an Assistant and Primary Examiner in the biotechnology group. While at the USPTO, an application she examined was selected as US Patent 5,000,000. Robin has given numerous talks and presentations relating to intellectual property legal issues, especially those relating to biotechnology. Current address: Pillsbury, Madison & Sutro, LLP, 1100 New York Avenue, N.W. Ninth Floor, East Tower, Washington, DC 20005-3918, USA. rteskin@PillsburyWinthrop.com

John Wooliams is Senior Principal Investigator, Roslin Institute, and Special Professor, University of Nottingham. As a researcher he has developed both operational and technical methods that are in widespread use in the livestock breeding industry. His main research emphasis has been on developing an integrated framework for understanding the dynamics of generating gain and loss of genetic variation. He has been closely involved in the work of the Food and Agriculture Organization of the United Nations to develop guidelines for sustainable breed development and for maintaining small populations at risk with a particular emphasis on developing countries. Current address: Roslin Institute (Edinburgh), Roslin EH25 9PS, UK. john.wooliams@bbsrc.ac.uk

Preface

> *I will pass through all your flock today, removing from there every speckled and spotted one, and every black one among the sheep, and the spotted and speckled among the goats. This will be my hire* (Genesis 30:32)

Ever since the time when individuals first domesticated animals for their own use and then later bred them for particular characteristics, there has been an effort to protect the breeder's developments and creations. Initially, this protection came in the form of development of breed societies and pedigrees. Breeders protected bloodlines by requiring others to buy purebred stock and to participate in the breeding organizations. In fact, the added cost of such endeavours and the added cost of buying genetically improved breeding stock can be thought of as a licensing fee for their hard work and developments.

Biotechnology breakthroughs in genomics have made it possible to identify genes, clone them and use them in far more sophisticated ways than the classical methods of selection and mating systems used by animal breeders. These new molecular biology discoveries and manipulation require huge investments and, like most situations in which science is pushed ahead by investments, some form of protection and financial gain is needed. Despite earlier attempts, this became a reality in 1980 when a series of legal challenges and successful appeals led to the US Supreme court allowing that inventions could include living matter (*Diamond vs. Chakrabarty*). The first patent for a multi-celled animal called the 'Leder' or 'Harvard' mouse was granted on 12 April 1988. Patents involving genes and their function have now become commonplace but the landscape for such developments is always changing.

Objections to patenting have been voiced by many, including consumers, food producers, ethicists and scientists. These objections include interference with nature, devaluing human life, animal welfare concerns, the 'polluting' of academic research and the undermining of agriculture and family farming. Other issues include concerns over the broad claims allowed with some patents, and the concerns that the research exemption (the use of a patented invention solely for the purpose of research or experimentation) will be disallowed. In no way is this book meant to convince the reader that patenting is both morally and ethically correct. Instead, it is meant to inform the reader on a variety of subjects related to patenting and animal breeding. Certainly, the concerns expressed by others will need to be continuously monitored. The patenting of genes and animals and how they are used to produce food, disease therapies, etc., will have a great impact on the way these are utilized in the genetic improvement of animals.

This is not the first such book devoted to animal patents; we direct the reader to the book *Animal Patents: the Legal, Economic and Social Issues*, edited by W.H. Lesser and published by Macmillan in 1989. It is our hope that *Intellectual Property Rights in Animal Breeding and Genetics* will be a useful and informative update on the ever-changing area of patents involving animals, with special emphasis on the role they play in the business of animal genetic improvement. The first chapter lays the foundation for the application of patenting to animals by defining patents and intellectual property. Chapter 2 provides a history of animal patenting and the inherent controversies that have ensued from policy development and execution. Chapter 3 is devoted to economic considerations and costs and benefits of patents. Chapter 4 discusses access to data and the conflict with protection of intellectual property. Chapter 5 examines protection of intellectual property on a global basis.

The next five chapters chronicle the impact of research and development (both publicly and privately) on aspects of patenting and intellectual property. Chapter 6 looks at the role that publicly funded universities play in the development of intellectual property, and the relationship between private industry funding and the preservation of intellectual property. A narrative on what has been learned from patenting plant material and how this might impact animal patenting is provided in Chapter 7. Chapter 8 looks at the legal ramifications of sequencing the human genome with applications to animals. Chapter 9 describes issues of intellectual property related to software development and the role of open source software. Ethical issues related to gene modification, biological diversity and patenting are discussed in Chapter 10.

Chapters 11–15 are actual case studies on discovery, intellectual property and patenting. These cover a wide variety of issues, starting with patenting of genetic markers for litter size in the pig and marbling in beef cattle. Additional chapters cover development of cloned livestock, development and commercialization of software for genetic improvement programmes in livestock and the controversy over a patent claim for a statistical methodology used extensively in genetic evaluation of livestock. Final thoughts from the editors of this book are provided in Chapter 16.

Finally, let us say this book is written with several groups in mind. It is designed to inform students, researchers, academics, industry personnel and others about patenting. It is the editors' hope that this book will be used both as a reference book and a classroom book. We encourage animal breeding and molecular and quantitative genetics groups and animal science departments to consider it as a text for discussions and seminar (this will certainly help book sales!). We are convinced that such discussions will lead to healthy and informative debates and contribute to the education process of all involved.

Max F. Rothschild
Scott Newman
1 March 2002

Acknowledgements

The development of a book for students and colleagues about the increasing role of patenting in animal breeding and genetics has been under consideration by MFR for several years. Thanks to the help of many people and institutions it is now a reality. The book was completed while MFR was on sabbatical leave at Sygen International in Berkeley, California. The financial support provided by Iowa State University and Sygen International for MFR is greatly appreciated. A special thanks is given to Gary Anderson for helping to provide use of library facilities at the University of California at Davis. The support of this endeavour by PIC USA for SN is also gratefully acknowledged.

A number of very important individuals have helped to make this book possible. First and foremost, MFR would like to thank Scott Newman, his friend and colleague, for agreeing to help co-edit this book and for his thoughts, discussions and many contributions to the development of this book. MFR and SN also very much want to thank all the contributors for their well-written and thoughtful chapters. These chapters were also improved from the comments of many reviewers. We thank Bees Butler, John Byatt, Greg Davies, Bruce Golden, Chris Haley, Donna Johns, Kenneth Kirkland, Dennis Knudson, Richard Lower, Heidi Nebel, Richard Nelson, Louis Ollivier, Peter Phillips, Jorge Piedrahita and George Wiggans for their very thoughtful and timely reviews. A special thanks goes to Richard L. Willham for his artful design of the cover. We also thank Denise Rothschild for help in preparing the index. The hard work and skilful help of Ms Rebecca Stubbs and Rachel Robinson, CAB *International*, in managing our efforts is also gratefully acknowledged.

The useful suggestions, support and friendship of our colleagues is gratefully acknowledged. Finally, the loving encouragement of our families in this endeavour is greatly appreciated.

Max Rothschild
Scott Newman
2002

1

Patents, Trade Secrets and Other Forms of Intellectual Property Rights

William Lesser

Department of Applied Economics and Management, Cornell University, 405 Warren Hall, Ithaca, NY 14853-7801, USA

Introduction

Origins of Intellectual Property Rights (IPRs)

The earliest form of IPRs has been traced by many scholars to Venice in the Middle Ages. Master craftsmen, it seems, were beset with competition from former apprentices. To control that competition, a law was passed which prohibited former apprentices from going into competition with licensed craftsmen for a time equivalent to the period of the apprenticeship (Walterscheid, 1994, *et seq.*). That period, about 20 years, is said to be the origin of the current 20-year patent duration. Certainly, craftsmen benefited from reduced competition. Apprentices benefited less so, but without the protection from competition, many may not have found masters willing to take on apprentices. Finally, the consuming public lost in the short run through higher prices, but with the longer-term assurance of trained master craftsmen available to produce the products, eventually benefited as well.

Purposes of IPRs

Economic justification

Over subsequent centuries, patent and related IPR laws have undergone multiple changes, but the essence of the public/private trade-off seen in long-ago Venice is still evident. Patents are temporary (typically 20 years[1]). The inventor benefits from that period of reduced competition and thereby has an incentive to invest in inventive activities. The public benefits from the existence of products and processes which otherwise may not have existed, or whose appearance would have been delayed. The justification recognizes the infringer, who does not have to recover the research and development (R&D) expenses of the inventive process, and can always sell at a lower price than the inventor. Hence, no one outside the public sector has an incentive to invest in new products which are easily copied, and technological advances stagnate. Patent laws restore that incentive through the legal prohibition on direct copying.

Contemporary patent law adds a further dimension, the requirement the invention be described. The resultant file of descriptions creates a public storehouse of technical information while helping assure competitive products can be produced soon after the patent expires. For these benefits, the public pays a higher price than would prevail in a competitive market. A higher price can come about only if supply is reduced so that the public has available less of the product than under fully competitive conditions.[2] Stated differently, product use diffuses more slowly. (For a more formal treatment, see, e.g. Maskus, 2000, Chapter 3.)

In addition to the mere existence of the novel product or process, the public benefits from the so-called spillover effect. Spillover recognizes that a private owner cannot capture all the benefits from a product. A large piece of property kept undeveloped by an owner not only benefits the owner, but the public, which enjoys the views and ecological benefits provided. Alternatively, an invention (transistor) can lead to additional, related inventions (inexpensive computers) that further benefit the public. Or the inventor, to meet unit sales and profit goals, may need to charge a lower price than many users would be willing and able to pay. Those users gain a direct benefit at the expense of the inventor.

The investment incentive is the principal justification for patents and related IPRs. Jefferson was well aware of this need, as he is often credited with inserting into the Constitution (Article 1, Sec. 8)[3] a requirement for the US Congress, 'To promote the progress of science and the useful arts, by securing for limited times to authors and inventors the exclusive rights to their respective writings and discoveries.' The Patent Act of 1790 was one of the early acts of the new Congress.

[1] From first application, so that the actual period is lessened by the examination period, which averages 2–3 years but can be longer. Some countries have provisions for protection prior to the patent issuing, or extensions of the term for particular products like pharmaceuticals.

[2] That is, demand curves are generally downward sloping so that higher prices imply reduced quantity.

[3] Walterscheid (2001) has presented evidence raising questions about Jefferson's role in the establishment of the US IPR system.

There, though, is another justification for patent-type laws, the provision of an incentive for firms to distribute products to other countries. Key to appreciating this aspect is a realization that intellectual property law is strictly national: protection must be held in each country to be effective. For example, a product protected in the USA has no protection in Canada or Mexico unless specifically sought and received there. Inventors should be concerned about unauthorized copying in those markets as well, and are typically hesitant to supply them without corresponding intellectual property protection. Hence, intellectual property protection can assist in the transfer of products and technologies.

Patents belong to a group of related rights known as industrial property rights, which also include Plant Breeders' Rights, trademarks and trade secrets (see also the later section). Trademarks function somewhat differently, but with a general similarity of intent. Trademarks allow owners to develop goodwill in a product with the knowledge that it will not be dissipated by unauthorized imitators. For the consumer, a trademark serves as a signal of product quality, for example a Coca Cola® or Burger King® hamburger have a consistency and standard taste that consumers expect. True Nike® athletic shoes are generally far higher quality than unauthorized copies. Thus, with trademarks too, both the owner of the trademark and the public benefit, at least to some degree.

Trademarks are related to Appellations of Origin, which are often applied to agricultural products like champagne. With the foundation that the particular growing conditions contribute to the quality (flavour) of the product, an appellation prevents others from appropriating the quality reputation of the 'true' product.

Copyrights fall into a separate class of protection mechanisms. Copyrights are applied to creative works like books and music, although important current uses include software. Copyrights are more protection by right, and involve fewer formal procedures. Indeed, in the USA, as in many countries, any creative work receives an implicit copyright, which can be signified by adding the © symbol. Registering the mark is beneficial should a legal action be required, but not strictly necessary.

Trade secret law imposes a penalty if secrets are acquired in an inappropriate way. The secret itself may be anything of value, such as customer lists or performance records. No formal process is required to achieve trade secret status, beyond a serious effort to keep the secret. In that way, a trade secret can conceivably be perpetual. The formula for Coca Cola® is a well-known trade secret.

In many cases, firms will achieve stronger (broader) protection by combining several forms of protection within a single product. For example, a patent might protect the product, but the least cost process for producing it is protected by a trade secret. Or a product is patented while the name is trademarked.

Moral justification

A second justification for IPRs is a moral, or perhaps more correctly, personal rights justification. This concept, often couched in the individualism philosophy of the Enlightenment, stresses that creation (whether artistic or technical) is from the self and hence should be granted the same rights as any personal property. IPRs and personal rights are aspects of the self and hence should be granted as a right (see Thompson, 1997, Chapter 8). Here, the economic justification will be stressed, as it is generally. The moral right concept, though, does continue in the idea of benefit sharing (see later sections) as well as requirements in most patent laws that the inventor(s) be identified by name. Perhaps the inventor does not own the invention (employment contracts, for example, typically make any invention in the course of work the property of the employer), but he/she/they at a minimum must be identified on the patent as the inventor.

Forms of IPRs Applicable to Animals

The preceding section laid out the general justification for IPRs, and identified in broad terms the major forms. This section contains more detailed descriptions of the operation of the several systems potentially applicable to animals. Greatest attention is given to patents.

Patents

Patents are intended as an incentive and reward for developing something new and useful, but only if the development is a non-trivial extension. The system has developed specific terms and interpretations for these concepts.

Non-obviousness or inventive step

The invention should not be obvious to someone 'skilled in the art'. This is the non-trivial requirement.

Novelty

The invention must be new in the sense of not having been previously known through publication or public use. The USA uses a system with a 1-year 'grace period', which means simply that the invention can have been revealed up to 1 year prior to applying for a patent. Most countries do not allow any prior revealing of the invention, in what is known as absolute novelty. Absolute novelty is particularly difficult for public sector researchers, whose work often requires publishing results.

Utility

Utility requires that the invention must serve some identified purpose. This does *not* say the invention must be practical, as patent offices do not judge practicality. Rather, the requirement is that a non-trivial use be identified.[4] This is an important component of the system, for it led to the rejection of applications for human genes early in the Human Genome Project. Those applications read to a gene construct only, with no specific knowledge of a use for it. Hence, the rejection. This is discussed further in Chapter 7.

Many patent offices, particularly those in developed countries, have employees who evaluate applications on these criteria compared to standards. This is known as an examination system. Some, however, merely register applications based on the claims of the applicant (registration systems). Registration is quick and inexpensive, and does not require technically trained personnel – all advantages. Presently, the US examination system requires about 2 years from application to issuance, but that may be longer for complex technologies. However, the validity of registered patents is not known unless and until there is a court challenge, which means many invalid patents may be on the books. For examination systems, an issued patent has a 'presumption of validity'. In practice, in the USA, up to 1% of patents are overturned in court challenges annually.

Issued patents are all of the same general type, though that masks some large differences. One is the *scope* of an individual patent, a wider scope means a related product must be more different not to be infringing. The scope of the first animal patent, the 'Harvard mouse', is quite broad, reading in the first claim, 'a transgenic non-human mammal' which exhibits the identified trait. Not until claim 12 is the non-human mammal limited to 'being a mouse'. Often, patents in new technical fields are broad, narrowing as the techniques become more standardized. For example, US animal patents number 3 and 4 claimed only 'a transgenic mouse' (Woessner, 2001). Typically, a person knowledgeable in the field is needed to determine the scope of a patent.

A second distinction is made as to whether the patent may be referred to as *per se*, process, product by process, or dependent. These are not terms which will appear on the patent, but can be inferred from a close reading. For example, a claim which reads 'a means of' is usually a patent for a process. A product by process patent covers the product only if it is produced in the described way. A different production technique lies outside the patent. The concepts of *per se* and dependent patents can be explained by reference to rBST, the compound used on injection to increase milk production from dairy cows. The initial (prod-

[4] That is, a gene construct cannot serve merely as a probe for self-identification, nor can a genetically engineered mouse be only 'snake food'.

uct by process) patent referred to that trait or utility. However, because it was the first, the patent owner had rights to any subsequently developed uses. One was identified: the use, early in the life of a heifer, leading to udder growth and greater milk production potential. That later patent, though, is dependent on the first one (if produced using the patented process) so cannot be used without the permission of that patent owner. The owner of the initial patent, for his or her part, also cannot use the improvement without permission. Often, what could be an impasse is resolved with a cross-licence between the two owners.

The matter of prior rights with living organisms, including animals, is critical, and is discussed in a further section.

Plant Breeders' Rights (PBRs)

PBRs is a special purpose system only for plants. In most countries, PBRs are administered by an office of the agriculture department, rather than by a separate patent office. Technically, it is a form of *sui generis* or special-purpose system, because of the exclusive focus on plants. PBRs are quite recent, with origins in the 1930s and 1940s. UPOV (the French acronym for the International Convention for the Protection of New Varieties of Plants), the international convention, was established only in 1961.

PBRs replace utility, non-obviousness and novelty for patent systems (see above) with *uniformity*, *stability*, *distinctness* and *novelty* (abbreviated as DUS). Novelty is the same concept as relates to patents, although the specific terms and use conditions differ. Uniformity and stability are technical requirements that assure the protected variety is identifiable after repeated multiplications. Clearly, if the protected product (variety) cannot be definitively identified, the system breaks down. For the future, genetic markers or a related technology may be used, but that is not the case at present.

Distinctness is akin to non-obviousness for patents. Distinctness may be claimed in one of a number of traits, such as disease resistance or flower colour. Most countries actually grow out varieties and measure the difference from a base or reference variety. In the USA, the claims of the applicant are generally accepted, akin to a registration system. Moreover, the US Plant Variety Protection Office (PBRs is strictly an informal if commonly used term) allows distinctness for traits of no practical merit. Hence, PBRs in the USA are generally considered to be weaker than in other UPOV-member countries (Lesser, 1987).

PBRs have other components that translate to providing weaker protection than patents. These are the so-called *Breeders' Rights* and *Farmers' Privilege*. Breeders' Rights specifically allow breeders to use protected varieties in a breeding programme without the permission of the variety owner. For patented products, this is known as the *research exemption*. However, because the research exemption is not statutory (written in the law), it is a matter of court interpretation, the focus of which is still under debate by scholars and practitioners. The Farmers' Privilege expressly allows farmers to retain the crop for use as a seed source for a subsequent crop. The farmer, however, may not sell the seed or share it with other farmers. The most recent UPOV Act (1991) makes the Farmers Privilege a national option, but most countries (including the USA) are choosing to allow it.

PBRs are intended to protect the entire plant, as well as its propagating parts (seeds, etc.). This is in line with the specific focus of preventing direct competition for that variety. A weakness became apparent when overseas growers (where PBRs were unavailable) shipped plant parts like flower blossoms into a protected market. That was not illegal until the most recent version of PBR legislation was adopted, extending protection to plant parts, and potentially to products of those parts (say flour from a protected wheat variety).

For biotechnological applications, though, the current scope of protection is insufficient. Breeders can simply use cross-breeding or other means to transfer a genetically engineered gene construct from one (protected) to another (unprotected) variety. The Breeders' Rights exemption makes such a transfer perfectly legal. A patented plant would provide protection from such a gene transfer, for the patent would read to the genetically engineered trait so protection would not be voided through the breeding process. Moreover, the breeding process itself would be a likely infringement. Conversely, PBRs covering the variety and a patent for the gene construct provide protection similar to a patented plant.

PBR-like systems (if adopted) have two advantages for animal breeders. They are simpler and less costly to secure, and hence appropriate for sequential improvements like livestock breeding. Second, many countries have specifically prohibited patents for plants or animals, but are allowing PBR systems. However, a workable system requires identifying the target population, in short, an operable definition for a 'variety'. That exists for plants, but the relatively small number of progeny and inherent heterogeneity of an animal line (with the possible exception of inbred experimental lines) makes a direct extension of PBRs to animals inappropriate. One proposal for an 'Animal Breeders' Rights' system is described in a later section.

Trade secrets

Since the 1930s, maize breeders have used trade secrets to protect first generation (F1) hybrids. Since hybrids do not reproduce true-to-type, the sale of seed does not allow the duplication of the seed. Breeders need only to protect the pure lines and crossings used to gain *de facto* protection of the resultant hybrid. Hence, breeders typically fence their pure line fields and, of course, do not reveal the crossing sequence.

For agricultural livestock, a similar approach is used by poultry and pig breeders. There, multiple crosses of pure lines produce 'synthetics' with characteristics similar to plant hybrids. Poultry breeders keep tight control of their pure lines, and release grandparent and even parent stock only under careful supervision. Of course, this strategy is not applicable for traditionally cross-bred animals but is now being employed by pig genetics companies with their larger customers.

Issues with Protecting Living Organisms

The preceding section describes the general considerations for protecting inventions. Living organisms, though, bring to bear several additional considerations. This can be inferred from the general recentness of extensions of protection to living matter. An early example is a patent granted to Pasteur in the 1870s for a yeast strain, but the Patent Office seemingly considered the product to be inanimate rather than animate. The initial specialized law was the (US) Plant Patent Act of 1930 (subsequently incorporated into the Patent Act). The Plant Patent Act applies only to asexually propagated plants (cuttings, bulbs and tubers, but not potatoes). Asexually propagated materials are stable, which overcame congressional uncertainty about the identification of protected materials over repeated regeneration. The exclusion of potatoes was apparently for concern about the implications for food prices, although fruit trees and berries were and are protectable. Flower breeders are major users of plant patents for both roses (propagated by cuttings) and bulb-reproduced varieties.

Subsequently, European interests took the lead in protecting plant varieties, beginning in the 1940s. The first international convention, UPOV, was adopted in 1961. The USA did not adopt a version, the Plant Variety Protection Act (PVPA), until 1970, amended in 1980. The US version excluded F1 hybrids; no specific justification has been given, but the Congress appeared to rely on the 'trade secret' protection afforded by hybrids (see previous section). The 1994 (the most recent) amendment dropped the exclusion of protection for hybrids. Chapter 7 details experiences under the PVPA.

Of major significance was the 1980 Supreme Court decision in *Chakrabarty*. While the particular decision related to an engineered microorganism (for more rapid decomposition of spilled crude oil), the decision was made in sweeping terms. The Court declared, 'anything under the sun that is made by man' is patentable subject matter. Shortly thereafter, in 1985, the Patent and Trademark Office (PTO) in an internal decision, *Ex parte Hibberd*, declared (higher) plants to be patentable subject matter. 'Patentable subject matter' simply means the application cannot be rejected solely because of the area of application.

Two years thereafter, in another internal decision in *Ex Parte Allen* (1987), the PTO declared higher animals to be patentable subject matter. Interestingly in that case, the underlying application for a 'polyploidal' oyster was rejected for insufficient non-obviousness. The first animal patent did not issue until April 1988, for the 'Harvard mouse', a laboratory research specimen. The number of animal patents continues to rise rapidly, with about 45 identified as 'mammals' (patent class 800/14) alone for 1995–2001 vs. none in earlier periods. Over 100 patents for transgenic mice have been issued. More are undoubtedly in the pipeline (until very recently, the PTO did not publish patents unless and until they issued). Most are laboratory specimens for use as human disease models, followed by livestock transformed to produce valuable compounds like pharmaceuticals (Woessner, 2001). The reason for the general absence of, say, agricultural livestock is not entirely clear. It may be due to concerns about an adverse public reaction, or the practical issues of identifying the patented animal. Back in 1987, there was quite a negative public response, to which the PTO responded with a self-imposed moratorium. The Congress threatened a moratorium, but one was never imposed.

Interestingly, in 2001 there has been a challenge heard by the Supreme Court to the provision of patents for plants (Pioneer Hi-Bred vs. J.E.M. Ag Supply). The challenge began (in typical fashion) as a defence for an infringement charge, declaring there was no infringement for the patent was invalid. The reason was that the Congress passed the PVPA to protect plants and hence did not intend to allow for the extension of patents. The Supreme Court decision was eagerly awaited, because the *Chakrabarty* decision was very specific, and the Court rarely undoes such a definitive decision. Yet, it is being asked, why then would the Supreme Court take time to reconfirm a clear (if 5 to 4) decision except perhaps to confirm the right of Congress to establish policy (Rories, 2001)? In any case, patents for animals would not be directly affected, for there is no *sui generis* system for animals. Opponents could, of course, take any change as a signal of an opportunity to oppose patents for living organisms more broadly, as many indeed do presently. In the end, the Supreme Court upheld the PTO practice of patenting plants.

Internationally, the status of patented animals (and plants) is far from clear (see Chapter 5). The issue goes back to the European Patent Convention (1973) which, in the infamous Article 53(b), prohibited patents for 'plant or animal varieties or essentially biological processes for the production of plants or animals'. With an origin dating to an earlier and simpler era, it is, with the advent of the new biotechnology, no longer easily determinable what is an 'essentially biological process'. This issue has been debated and legislated numerous times in Europe, with the general agreement that a 'variety' applies to the fixed form. Any innovation (like a genetically modified one) that is applicable to multiple varieties is generally considered to be patentable. Similar reasoning can be applied to animals, although there is no complete resolution at this time. Article 53(a) allows for excluding patents for 'ordre public' and morality. Some attempts, generally unsuccessful, have been made to identify animal patenting as immoral (see Grubb, 1999, Chapter 14).

This European Patent Convention wording is also significant because it has carried through in modified form into the TRIPs (Trade Related Aspects of Intellectual Property Rights) of the 1994 World Trade Organization. Signatories (all 144 countries, currently) have agreed to adopt certain minimum forms of intellectual property protection. Some wording is straightforward, the provision of trade secret legislation is an example. The area of plants and animals, though, remains ambiguous, for, in Article 27.3(b), countries have the option of excluding from patentability 'plants and animals other than micro-organisms, and essentially biological processes for the production of plants or animals. However, Members shall provide for the protection of plant varieties either by patents or by an effective *sui generis* system or by any combination thereof.' While the situation in some developed countries is ambiguous regarding the patenting of animals, as is noted above, virtually all developing countries are choosing to prohibit patents for plants and animals. In this case, the bar to the patenting of animals is more absolute than under the European Patent Convention for the wording reads to 'animals', not 'animal varieties'. In consequence, it will likely be some time, at best, before animals are patentable in developing countries.

Effects on Research

Numerous reports have been made of the deleterious consequences of IPRs on research, particularly in the public sector (see, e.g. Eisenberg, 1989). A typical case would involve a researcher who acquired some component of the final product through a Material Transfer Agreement (MTA), which provided for research use only. Once the final product is ready for commercialization, the researcher is required to negotiate terms with the owners of the input materials. Negotiations are at minimum lengthy and costly, and at worst may lead to no practical agreement. The researcher and public are then denied the product. Negotiating commercialization agreements prior to the identification of the product is conceptually possible and would avoid the final 'hold-up' problem, but with so many projects never leading to practical products, negotiating in anticipation would place a major time and cost burden on the public sector.

The potential for this kind of problem should not be trivialized. For one plant product, 'GoldenRice', investigators counted between 0 and 44, depending on the country, patented components (Kryder *et al.*, 2000). In truth, MTA are contracts, so technically not IPRs, but IPRs can and, as seen in the GoldenRice case, do cause identical kinds of problems. A search of claims of ownership of components of a product is known as a 'freedom to operate' review.

The private sector, when faced with similar issues, may cross-licence technologies, something known as 'bargaining chips' in the literature. Alternatively, a firm may be less than scrupulous in determining freedom to operate with the assumption/expectation that the owners will not vigorously pursue their ownership rights (von Hippel, 1998). Infringement suits can easily cost into the millions, although most are settled out of court for just that reason. Sometimes public sector researchers imprudently dismiss the possibility of prosecution, if only because of the bad publicity of a private firm suing, say, a university. Publicity may indeed be a deterrent to formal court action, but perhaps the bigger issue for researchers is reduced access to research materials in the future. Most public sector researchers cannot conduct frontier level work without some materials from the private sector. The goodwill required would be gravely affected by a careless violation of a patent or MTA.

Most of the cases to date have been for plants, but the situation for animals is very similar. There is at this time no identified way to avoid such hold-ups, but careful record keeping at least tracks the size and source of the possible problems.

Some patented products seek, as part of the royalty payment, some rights to the resulting products. These requests might apply to the use of the 'gene gun' for creating the modification, or the use of a particular laboratory specimen. These agreements, again, apply only and if there is a commercializable product. From an economic perspective, these arrangements reduce the cost of research by shifting some of the payment (and risk) to the product stage when there will be an income flow. However, the independence and flexibility of the inventor is restricted when commercialization deals are arranged.

Issues in Extension of IPRs to Animals

Conceptually, there are no limits to the extension of IPRs to animals. Indeed, to the extent they have been applied to date, no obvious problems have appeared. There have been some complaints about the high prices of patented laboratory animal models. Such complaints, though, must be balanced against the recognition that the specimen might not have existed had it not been for the existence of the patent option. Once patented animals are broadly distributed outside the confines of the laboratory, additional practical issues will arise. This can be most readily appreciated by agricultural livestock, although similar concerns would potentially apply to other classes of animals, such as pets and racehorses. The most extreme case is cattle, for, due to their slow reproductive rate, any trait must be disseminated into the herd by breeding.

One attribute of slow dissemination, recognizing that traits will not be inherited 100% if only one parent is a carrier, is incomplete knowledge of which individual carries which trait. (This scenario assumes no visible 'marker' is associated, and that no rapid and inexpensive testing mechanism exists.) The issue then arises: who is liable to pay a royalty, and when in the production cycle? The patent owner, of course, prefers payment early, say for the calf. Yet it will not be possible, then, to identify which traits are being carried (for over time several will accumulate). Plus, the producer will be asked to pay for a benefit that will not be realized until slaughter, months or years into the future. Significant, too, is the difficulty of identifying where and when calves are produced in the dispersed cow–calf sector. For those reasons, Lesser (1989) has proposed a royalty collection system for agricultural livestock with the packing plant serving as the focal point.

Several benefits exist for having packing plants serve the royalty collection process:

- A limited number of plants eventually process essentially the entire herd (presently the largest four firms process 82% of cattle in the USA, for example),
- A sample can be taken to determine the distribution of the traits in the herd, with royalty claims based on the herd average,
- Charges are placed when revenue is realized, reducing the cash flow impact on the livestock producer, and
- Multiple trait owners will be made aware of the aggregated royalty charge, reducing the likelihood of individual requests pyramiding into an uneconomical total.

Inequities and complexities nonetheless remain. Notably, with producers paying on the basis of the herd average distribution of traits, some will overpay and others underpay for their individual animals. Inventors must wait for extended periods until collecting royalties, creating a cash-flow problem for at least the small firms, and packing plants may be reluctant to take on the role of a collection agent for multiple inventors. Some of these complexities will be resolved when a rapid system, perhaps using DNA markers, is developed for identifying the existence of patented traits in individual animals. Yet, whichever system eventually emerges, considerable coordination will be required of a type not previously needed and for which there is no existing agent.

Lesser (1993) also attempted to identify a workable *sui generis* system for protecting animals. Such a PBR-like system would be useful for protecting the output of routine breeding that may not qualify for the non-obviousness requirement for a patent (see previous section). The relatively low reproductive rate of animals (compared to plants) and the absence of an operational designation of an animal 'variety', however, exclude a simple extension of the concepts of PBR systems. Lesser's (1993) proposal, therefore, combines components of trademark systems under which the genetic materials are not owned, but their value is reflected in the sires' identification. Consider for example an application to racehorses. The seller of a foal would not pay directly for 'Man of War's' genes, but would be assessed a pre-set fee if Man of War were identified as a sire. Additional proposals/concepts are required, but the current absence of an animal variety rights system, as a complement to patents, is a disadvantage to attracting additional investment to animal breeding.

The enhancement of any animal line in the final analysis depends on the genetic diversity available to breeders and scientists. For the major food grains (wheat, rice, etc.), gene banks contain somewhere from 55 to 80% of traditional crop varieties and 87% of domesticated animals. For wild species, however, the estimate for crops is from 1 to 70%; estimates for animals are generally not available (Heywood, 1995, Chapters 3 and 13). Any collection of genetic resources requires a careful consideration of ownership, access and use rights, and any benefit sharing with the communities associated with the development of a 'breed', including associated indigenous knowledge. For plant genetic resources, these matters have been under development by the Food and Agricultural Organization (FAO) of the UN for decades, principally under the International Undertaking (now Treaty) for Plant Genetic Resources. The voluminous details of those arrangements exceed the purpose here, but there is a clear association between the development of IPR and requests for the sharing of benefits perceived to flow from those rights. The animal sector should be forewarned of the practical and moral needs for such arrangements as property rights systems expand.

Conclusions

There is no conceptual reason why patents cannot be extended to animals. The relatively few cases (compared to other classes of patentable subject matter) where this has been allowed and done have been applied smoothly. However, most countries worldwide (IPR protection being strictly national) have explicitly excluded patents for animals, while others, by means of complex interpretations of legal language, have made applications uncertain. Hence, it will at best be many years before animals are treated equivalently with other applications before the patent system. Moral objections to the patenting of animals, while not presently a major legal impediment, do and will affect the use of patents and acceptance of associated products by sectors of the public. Moreover, there is no existing system anywhere which parallels PBRs for traditionally bred animals. The likely consequence overall is less investment in the animal sector than would occur with stronger and more universally accepted property rights protection.

If patents are applied to animals, such as agricultural livestock, which are dispersed broadly, a number of largely practical problems will arise. These include matters such as identifying which traits are present in individual animals, and determining appropriate royalty rates recognizing the potential of pyramiding rates to supra commercial levels. Finally, attention needs be given to the collection and preservation of animal genetic resources, including the benefit-sharing issues that arise once value is associated. While these are significant issues individually and collectively, the animal sector has the advantage of observing data collected over generations and institutional modifications with IPR protection of plants, and benefiting from the understandings gained.

Acknowledgements

This work was supported in part by an USDA IFAFS grant. The author thanks an anonymous reviewer for multiple helpful comments.

References

Eisenberg, R.S. (1989) Patents and the progress of science: exclusive rights and experimental use. *University of Chicago Law Review* 56, 1017–1086.

Grubb, P.W. (1999) *Patents for Chemicals, Pharmaceuticals and Biotechnology.* Clarendon Press, Oxford.

Heywood, V.H. (gen. ed.) (1995) *Global Biodiversity Assessment.* Cambridge University Press and UNEP, Cambridge.

Kryder, R.D., Kowalski, S.P. and Krattiger, A.F. (2000) The intellectual and technical property components of pro-vitamin A rice (*Golden*Rice). ISAAA, ISAAA Briefs 20-2000, Ithaca, New York.

Lesser, W. (1987) The impacts of seed patents. *North Central Journal of Agricultural Economics* 9, 37–48.

Lesser, W. (1989) Animal patents in the USA – are the concerns justified? *Patent World* 13, 18–23.

Lesser, W. (1993) Animal variety protection: a proposal for a US model law. *Journal of Patent and Trademark Office Society* 76, 697–715.

Maskus, K.E. (2000) *Intellectual Property Rights in the Global Economy.* Institute for International Economics, Washington, DC.

Rories, C.C.P. (2001) Does the U.S.P.T.O. have authority to grant patents for novel varieties of sexually reproducing plants? *Journal of Patent and Trademark Office Society* 83, 737–758.

Thompson, P.B. (1997) *Food Biotechnology in Ethical Perspective.* Blackie, London.

von Hippel, E. (1998) *The Sources of Invention.* Oxford University Press, Oxford.

Walterscheid, E.C. (1994) The early evolution of the United States Patent Law: antecedents (Part I). *Journal of Patent and Trademark Office Society* 75, 398–426.

Walterscheid, E.C. (2001) The nature of the intellectual property clause: a study in historical perspective (Part 1). *Journal of Patent and Trademark Office Society* 83, 683–690.

Woessner, W.D. (2001) The evolution of patents on life – transgenic animals, clones and stem cells. *Journal of Patent and Trademark Office Society* 83, 830–844.

2

The Advent of Animal Patents: Innovation and Controversy in the Engineering and Ownership of Life

Daniel J. Kevles

Department of History, Yale University, PO Box 208324, New Haven, CT 06520-8324, USA

Introduction

In April 1987, an appeals board in the US Patent and Trademarks Office (PTO) (*in Ex Parte Allen*) set a precedent in the history of the American patent system by announcing that patents could be obtained on living animals. The board's adventurousness gratified biotechnologists, but it also disquieted wildlife advocates, who warned that it might threaten the environment, and some farm groups, who feared that it would work harmful economic effects in agriculture. Moreover, the World Council of Churches attacked animal patenting, declaring that it 'removes the distinction between life and nonlife' and admonishing that 'the gift of life from God ... should not be regarded as if it were a chemical product' (*Ex Parte Allen*, 1987; US Congress, 1987b, pp. 108–109; US Congress, 1989, pp. 258–272). The dissenters, convinced that the action of the PTO raised profound issues of political economy and ethics, insisted that it had been high-handed in expanding the scope of patent protection to higher life forms on its own – that such a momentous change in patent policy should have required congressional action.

What is patentable, according to congressional statute, dates back to the patent law of 1793, which declared, in language written by Thomas Jefferson, that patents could be obtained for 'any new and useful art, machine, manufacture, or composition of matter, or any new or useful improvement thereof.' Jefferson's phrasing remained – and remains – at the core of the US patent code, except for the 18th-century word 'art', which was replaced in a 1952 Congressional overhaul of patent law by the word 'process' (Bugbee, 1967; Machlup, 1968). The code said nothing about patenting living organisms, but a key precedent discouraging it was established in 1889, when the US Commissioner of Patents laid down what came to be known as the 'product-of-nature' doctrine – that while processes devised to extract what is found in nature can be patented, objects discovered there cannot. They are not inventions, nor can they as a class be made anyone's exclusive property. In the Plant Patent Act of 1930, Congress had granted patentability to one class of living products: plants that could be reproduced asexually (Bugos and Kevles, 1992).

Patents for Microorganisms: *Diamond vs. Chakrabarty*

There was no other extension of patent law to vital entities for 40 years, but then along came Ananda Chakrabarty, a biochemist at the General Electric Company, who in 1972, having bioengineered a bacterium to consume oil slicks, filed for a patent on the living, altered bacterium. The US PTO denied him a patent, arguing that no patent could be issued on a living organism, not least because it was a product of nature. Chakrabarty appealed his case through the courts, and at the end of 1979 it reached the US Supreme Court under the rubric of *Diamond vs. Chakrabarty*, in recognition of the fact that the position of the patent office was formally defended by Sidney Diamond, the current Patent Commissioner (Kevles, 1994).

By the time the case arrived at the court, it had become charged with the social and economic stakes that surrounded the swiftly accelerating commercialization of molecular biology through biotechnology. The trend to commercialization had been technically enabled by the advent of what was called 'recombinant DNA' – techniques devised in the early 1970s that permitted scientists to snip single genes from the genome of one organism and insert them into the genome of another. Companies like the adventurous start-up Genentech were being founded at a rapid pace, while major pharmaceutical firms, as well as several oil and chemical giants, were initiating molecular biological research programmes of their own, letting research contracts to the start-ups, and even obtaining an equity interest in some of them (Kevles, 1994).

Chakrabarty had not used the technique of recombinant DNA to engineer his oil-eating bacterium, but the issue that his case raised – the patentability of living organisms – spoke directly to the rapidly increasing stake in biotechnology patents. Ten *amicus* briefs were filed in the case. Most supported Chakrabarty and came from the nascent biotechnology industry, pharmaceutical manufacturers, patent lawyers, professional societies in biomedicine, and universities expecting to patent discoveries made in their laboratories of biology (Kevles, 1994).

On 16 June 1980, the Court held, by the slim margin of five to four, that whether the invention was alive or dead was irrelevant to its patentability; that the bacterium was not a product of nature, that it was a product of Chakrabarty and hence deserved a patent. Chief Justice Warren Burger delivered the majority opinion, enthusing over the broad language that Thomas Jefferson had written into the patent law of 1793, calling it expressive of its author's 'philosophy that "ingenuity should receive a liberal encouragement"'. Rejecting the contentions of the PTO, he found that the patent code as written was capacious enough to cover living microorganisms. Chakrabarty's bugs were new compositions of matter, the product of his ingenuity, not of nature's. As such, they were patentable under existing law (*Diamond vs. Chakrabarty*, 1980).

After the *Chakrabarty* ruling, several critics insisted that the decision appeared to leave no legal obstacle to the patenting of higher forms of life, including animals and, possibly, human beings. In fact, in the years between the *Chakrabarty* and the *Allen* decisions, a number of biologists began genetically engineering animals primarily for research purposes – for example, to study how cells differentiate as an animal develops from a newly fertilized egg or to explore the genetic dynamics of cancer. However, their efforts produced animals that were new compositions of matter and that in some cases had commercial and, hence, patentable possibilities.

The Genetic Engineering of Animals

The genetic engineering was accomplished by inserting foreign genetic material into an animal's genome using a method that had been recently devised independently in several laboratories. The fundamental step in the process was to introduce foreign DNA obtained with recombinant techniques into a newly fertilized mammalian egg. The immigrant DNA could integrate into and then proliferate with the creature's native genome, eventually finding its way into every cell of the grown animal, including its sex cells. When the animal reproduced, the DNA would be transmitted to some fraction of its progeny, automatically supplying a large number of such genetically transformed animals. The efficacy of the method was demonstrated in 1980 at Yale University by Jon W. Gordon and Frank H. Ruddle, who declared their results to mean that 'genetic transformation can be extended to whole mammalian organisms at a very early stage in their development' (Gordon *et al.*, 1980; Brinster and Palmiter, 1986).

By then, engineering animals with foreign genes – 'transgenic animals', to use the term that Gordon and Ruddle soon coined – had begun to shape the collaborative research of Ralph Brinster, a veterinary biologist at the University of Pennsylvania, and Richard Palmiter, a molecular biologist at the University of Washington. They first introduced a test construct of foreign DNA that comprised a gene attached to a sequence of DNA taken from mice that would promote and regulate the gene's expression. The promoter/regulator gene did its job, and they obtained expression of the gene in the recipient mouse (Brinster and Palmiter, 1986).

Brinster and Palmiter then decided to try their method of gene injection to correct dwarfism in a strain of mice that resulted from a lack of growth hormone. The strategy was to introduce into the mice the gene for rat growth hormone (rGH) attached to the gene for the protein metallothionein (MT), the same one that had worked in the test construct. MT regulates the level of heavy metals in the body; its gene is abundantly expressed in many organs, especially the liver. The ingestion of, say, too much zinc would stimulate the production of MT, which would then tie up the zinc. Palmiter proposed to exploit that process to control the expression of the growth hormone gene in the transgenic mice. By regulating the amount of metals such as zinc in their diet, he expected to control the action of the MT gene and, thus, of the growth-hormone gene (Brinster and Palmiter, 1986; Palmiter, 1990).

However, before introducing the rGH–MT combination into dwarf mice, Brinster and Palmiter thought to inject it first into the newly fertilized eggs of normal mice to check that it was properly constructed, that it would integrate into the genomes of the mice, and that it would express itself in their livers after they were born. Palmiter recalled that a couple of months after the mice were made, Brinster telephoned to report, 'They're growing larger than normal!' A number were 20–40% larger. Some were almost twice as big (Palmiter, 1990). In December 1982, Brinster and Palmiter reported their results in *Nature,* which gave their findings prominent coverage, including a cover picture of a mouse made huge by the growth hormone gene crouched next to one of normal size. They then achieved even better results using human growth hormone instead of the rat variety and, in November 1982 published their findings in *Science*, which heralded their article with a dramatic cover illustration comparable to *Nature*'s (Palmiter *et al.*, 1982, 1983).

The graphic demonstration of what might be achieved with transgenic animals excited many biologists, but perhaps none more than Brinster and Palmiter themselves. Brinster, the product of a farming family, had long been concerned with the improvement of agricultural animals. Early in 1982, he had established a collaborative effort to produce transgenic animals with scientists at the US Agricultural Research Service Center in Beltsville, Maryland. The group experimented with transgenic sheep for about a year but failed in most of them to obtain integration of the transgene into their genomes. Then Brinster began working with pigs. By that time, he and Palmiter had achieved their spectacular success with mice. Palmiter remembered that because the mouse grew after the insertion of a growth-hormone gene, 'we now expected everything to grow.' He and Brinster reasoned that pigs, made transgenic by injection of such a gene, might grow to twice current market size or grow to market size in half the time[1] (Brinster and Palmiter, 1986; R. Brinster, 1991, personal communication).

Working with a Beltsville scientist named Vernon Pursel, Brinster and Palmiter used a transgenic construct for the pigs that coupled the MT promoter gene to the gene for human growth hormone. Some of the resulting transgenic pigs grew bigger than normal ones; they also were less fatty and converted feed into meat more efficiently. However, they suffered from various afflictions, including gastric ulcers, kidney disease, arthritis, lameness, lethargy and a tendency to injury when they moved. They also lacked libidinal energy, a condition that dampened expectations that, once created, they would happily reproduce themselves (Pursel *et al.*, 1989a, b).

The Beltsville experiments indicated that transgenic transformation that worked well in one species, the mouse, would not necessarily work well in another, pigs. Nevertheless, Brinster, Palmiter and Pursel continued their experiments with pigs, trying to find a promoter that would work. Scientists in universities and biotechnology firms embarked on transgenic research programmes with other farm animals. They hoped to engineer chickens with higher resistance to disease, for example, or sheep that produced more wool, or cows that provided more meat (Jones, 1987; Saltus, 1988; *Toronto Star*, 1988).

Some also pursued what came to be called 'molecular farming' – genetically modifying animals so that they manufactured valuable proteins. Genentech had obtained its first commercially significant product, human insulin, by inserting the gene for human insulin into bacteria, which then manufactured its constituent elements, but animals were far more efficient producers of mammalian proteins. The gene for, say, Factor 9, a blood clotting agent that haemophiliacs lack, could not be produced in bacteria; but it might be inserted into a cow's genome and harvested from its blood (Jones, 1987; Saltus, 1988; Palmiter, 1990). In all, molecular farming appeared to offer a means of turning common animals into factories for the production of valuable human proteins that were otherwise difficult and expensive to obtain, if they could be obtained at all.

The First Animal Patent

While Brinster and Palmiter were pursuing the line of research that led them to Beltsville, still other scientists were creating transgenic laboratory animals that would serve as models for the study of genetically

[1] Brinster and Palmiter also introduced growth-hormone gene into rabbits.

based diseases. At Harvard University, Philip Leder and his postdoctoral collaborator Tim Stewart developed a transgenic mouse that was super-susceptible to breast cancer because it contained an oncogene, i.e., a tumour-causing gene, tied to a promoter that would activate the gene in the mammary glands. The work had not been done for the sake of devising a patentable product, but once it was accomplished, Leder recognized that it might have commercial possibilities. About the end of 1983, he brought his mice to the attention of the Office of Technology Licensing and Industry Sponsored Research, the recently established patents arm of the Harvard Medical School (Leder, 1988, 1991, personal communications).

Paul Clark, a patent attorney from the downtown Boston law firm of Fish and Richardson, Harvard's principal outside patent counsel, saw that Leder's transgenic animals were, like the bacteria in *Chakrabarty*, new compositions of matter made by man. Thus, he later explained, 'It was hard for me to see any legal basis for excluding claims on animals' (Atkinson, 1990, Clark, 1991, personal communication). On 22 June 1984, on behalf of Harvard University, Clark filed an application for a patent on Leder and Stewart's invention, which he took to be any transgenic mammal, excluding human beings, containing in all its cells an activated oncogene that had been introduced into it – or an ancestor – at an embryonic stage.

The same year that Harvard filed for a patent on Leder's mouse, the process that resulted in *Ex Parte Allen* was initiated by the marine biologist Standish K. Allen and collaborators at the University of Washington. They applied for a patent on a version of *Crassostrea gigas*, a variety of the Pacific oyster, which they had improved by making it chromosomally triploid. The claim was partly for the triploidy process, which made the oyster more edible. However, it also covered the improved oyster as such, which of course challenged precedent.

The examiners in the US PTO denied the claim, holding that neither *Diamond vs. Chakrabarty* nor any other patent ruling authorized the grant of a patent on a higher animal, even if only an invertebrate. The examiners also found that the triploid oyster was not patentable on the technical grounds that the innovation was obvious to anyone schooled in the art of oyster breeding. Allen and his colleagues appealed to the Board of Patent Appeals and Interferences, of the US PTO. In *Ex Parte Allen*, the Board upheld the examiners on the point that obviousness of art disqualified the oyster for a patent, declaring that patents could in principle be granted on non-human animals (*Ex Parte Allen*, 1987).

The ruling in *Ex Parte Allen* did defuse one important public fear about law and biotechnology, by stipulating that human beings cannot be patented by reason of the 13th Amendment to the US Constitution. Since it outlawed slavery, it in effect prohibits one human being from holding a property right in another. Following *Ex Parte Allen*, the patent examiners granted Harvard University its claim on Leder and Stewart's mouse, thus issuing the first patent on an animal in the history of the world's patent systems. By mid-1989, some 76 applications for animal patents had been submitted to the patent office (US Congress, 1989, p. 12).

Resistance and the US Congress

During arguments in the *Chakrabarty* case, vigorous objection to patents on living organisms had come from the People's Business Commission (PBC), an activist group headed by Jeremy Rifkin, a social agitator and sleepless critic of biotechnology (Kevles, 1986). The PBC's dissent was partly economic – patents on living organisms would foster monopoly in vital areas such as the food industry. It was quasi-religious, too, holding that 'the essence of the matter' was that to permit patents on life was to imply that 'life has no "vital" or sacred property', that it was only 'an arrangement of chemicals, or mere "compositions of matter"' (PBC, 1979). In its ruling on the case, the Supreme Court majority took note of these and other apprehensions, observing that they 'present a gruesome parade of horribles' and 'that, at times, human ingenuity seems unable to control fully the forces it creates'. The majority observed, however, that genetic research with its attendant risks would likely proceed with or without patent protection for its products, and that neither legislative nor judicial fiat as to patentability would 'deter the scientific mind from probing into the unknown any more than Canute could command the tides' (*Diamond vs. Chakrabarty*, 1980).

In 1984, still determined to defend the integrity of living creatures, Jeremy Rifkin filed suit in federal district court to halt the experiments with pigs under way at Beltsville. Declaring them 'morally reprehensible,' Rifkin said it was 'shocking that the US government would condone the introduction of human genes into an animal (Rensberger, 1984; Russell, 1984). The suit failed, but the *Ex Parte Allen* ruling in 1987 induced further outcries from Rifkin, and opposition to transgenic research with higher animals gained leverage when about the same time reports were published in the American and Canadian press about the Beltsville experiments, including the health problems of the pigs (Jones, 1987; Saltus, 1988).

In mid-May 1987, Senator Mark Hatfield of Oregon, who sympathized with the animal rights movement, urged the Commissioner of Patents and Trademarks to impose a moratorium on the processing of animal patent claims until congress could develop a policy on the issue. He also got the Senate to pass a 2-month moratorium as an amendment to an appropriations bill (*New York Times*, 15 May 1987; US Congress,

1987b, p. 2). In the House, Congressman Robert Kastenmeier, of Wisconsin, chairman of the House Judiciary Subcommittee that dealt with patents, attempted to come to grips with the issue by holding hearings on animal patents in mid-1987 and again in September 1989 (US Congress, 1987b, 1989).

Kastenmeier, a left-of-centre Democrat, was thoughtful and forthright, his judgments well considered, his approach to issues 'cerebral', as one of his staff put it. He had been wondering since the advent of recombinant DNA in the early 1970s whether genetic engineering comprised a wonderful advance or raised a terrible spectre. He believed that Americans were divided about the issue; the closeness of the Supreme Court's ruling in the Chakrabarty case was indicative of the division. He appreciated the social activists' worries about animal patenting, taking the reports of 'new varieties of pig, rust colored and large' as reason enough to concern 'those who do not wish to see animals mistreated or monster forms created'. He was also sensitive to the economic implications of animal patenting (US Congress, 1987a; R. Kastenmeier, 1987, 1991, personal communications). His congressional district included farming communities, the University of Wisconsin and the agricultural biotechnology firms that were growing up around it.

Kastenmeier doubted the adequacy of *Ex Parte Allen*'s prohibition against patents on human beings: did it extend to human organs, genes, or other parts? He doubted, too, that the answers to such questions should be left entirely to the Commissioner of Patents. He thought that the extension of patent protection to higher organisms constituted a 'quantum leap', one that the PTO had been high-handed in making. In his view, the Constitution gives power in determining the scope of patents to the Congress, and Congress ought to exercise it. Amid the controversy *Ex Parte Allen* had aroused, he considered it imperative to probe the merits of animal patents before the issuance of such patents turned into a policy that was both broad and irreversible (Kastenmeier, 1987, 1991).

The proposed moratorium on the issuance of animal patents drew support from animal rights activists, clerics and environmental-minded critics of animal patents. At the hearings, John Hoyt, the president of the Humane Society of the USA, attacked animal patents on both pragmatic and principled grounds. Such patents provided incentives for the kind of genetic engineering that produced the suffering of the pigs at Beltsville. They reflected 'a human arrogance towards other living creatures' that denied 'the inherent sanctity of every unique being and the ... ecological and spiritual inter-connectedness of all life'. They also suggested 'that animals have no inherent value other than that which serves the end of human beings' (US Congress, 1987a, pp. 55–56, 1989, pp. 108–109, 115). In the view of a Lutheran bishop, genetic engineering as such was morally dubious and so was the reduction of life to mere material objects. John Barnes, a veterinarian representing the Alliance for Animals and the Federated Humane Society of Wisconsin, wondered whether the genetic engineering of animals might produce 'a Frankenstein effect', unleashing a dangerous creature that might 'inadvertently escape from laboratory isolation ... and threaten life on earth'. Barnes claimed that animal patenting would 'compromise the integrity of animal species and ultimately lead to the control of life forms by a few multinational corporations' (US Congress, 1987b, pp. 345–346, 351–352).

Congressman Charles Rose, of North Carolina, whose wife was an enthusiast of animal rights, appeared as a witness to warn that scientists were 'playing God' by putting human genes into animals. 'What would happen to the rodent community in America if [rats with human growth genes] got loose in our society?' Rose was unwilling 'to let the market place determine the future of the animal kingdom'. Neither was Congressman Benjamin Cardin, of Maryland, a member of the subcommittee. He wondered whether the PTO would have any trouble dealing with a two-headed animal – part dog, part cat – that was claimed 'as a useful invention ... for a sideshow and circus'. There might be great demand for such a creature because of 'a lack of good freaks today', he noted. In light of *Ex Parte Allen*, he figured, with obvious disapproval, that the PTO would have no problem granting a patent on it (US Congress, 1987b, pp. 110–111, 31–32).

Some farm spokespeople chimed in with ethical and environmental objections to animal patenting. Gervase Heffner, of the National Farmers Organization in Wisconsin, said he was 'real concerned' about how in, say, 15 years, would farmers 'market that hog using human growth hormones? Where is that going to fit in the showcase in a store?' He declared that if one of those experimental pigs was 'put out into the market, I would stop eating pork' (US Congress, 1987b, pp. 313; see also pp. 329–333). But for all the grandiloquent and homely posturing that the ethical issues stimulated, far greater attention in the hearings went to the potential economic consequences of animal patents on farming.

Spokespersons for small farmers warned that the consequences would be disadvantageous to family farmers. Among them was Stewart Huber, representing the Wisconsin Farmers' Union, who leveled multiple complaints against animal patenting. Much of the research that undergirded the creation of potentially patentable animals had been paid for by the public; yet private corporations would, for the period of the patent, be gaining a monopoly property right in the genetically engineered animal, forcing consumers to pay twice for the same product. Allowing patents on genetically engineered animals would 'shift the profit motive for livestock improvement from the family farmers, who have used the classical breeding practices over the years, to the giant corporations which have the resources to use ... DNA research for their own benefit'. The process would have 'a chilling effect on traditional family lifestyle farms', speed vertical integration in farming, and make 'individual farmers ... wards of Wall Street in the biotechnology establishments'. Dealing with the progeny of patented animals would pose legal and financial burdens that would be both impractical and unjust. 'Farmer-to-farmer livestock sales ... could be construed to infringe on corporate patent rights, making the farmer subject to civil prosecution every time an animal was sold' (US Congress, 1987b, pp. 307–309).

What bothered small farmers did not, however, worry agribusiness. James Terrell, of the American Farm Bureau Federation (AFBF), told Kastenmeier's subcommittee that the Federation did not consider the ethical concerns important and that they should not be allowed to interfere with the incentives that patents provided for the engineering of better farm animals. He added that the AFBF did recognize that the disposition of rights in the progeny of the animals might pose problems but that they were not insuperable. The Federation, he stressed, did not fear monopoly pricing of prize animals. If producers regarded the patented animal as sufficiently valuable, they would pay for it; if not, not (US Congress, 1987b, pp. 124–127).

Strong defences of animal patenting came from other agribusiness witnesses as well as from academic scientists, and the biotechnology industry. They stressed that new farm technologies were essential to feed the world's growing population and were indispensable to American farmers in competing in the international market. Noting that the Japanese were taking steps to encourage their biotechnology industry, they warned that the USA had to guard against losing its biotechnological edge to Japan, as it had done in microelectronics. Practising scientists pointed out that patents were preferable to trade secrets, because they compelled the patent holder to disclose the details of the protected innovation, thus giving others the opportunity to improve upon it. Philip Leder explained the significance of Harvard's oncomouse for understanding breast cancer, suggesting that animal patents were necessary to attract industrial investment to advanced medical research that relied on animal models for developing diagnostics, therapies, and cures for human disease (US Congress, 1987b, pp. 114–135, 208–258, 320, 350; 1989, pp. 193–202).

The key witnesses for academic and industrial biotechnology were Tom Wagner, Winston Brill and Richard D. Godown. Wagner, a descendant of five generations of Ohio farmers, was the director of the Edison Animal Biotechnology Center, a state-supported consortium at Ohio University, his home institution in Athens. Brill, a biologist, was Vice President for Research and Development of the Agracetus Corporation, a biotechnology firm concerned with agriculture. Godown was President of the Industrial Biotechnology Association. They all adamantly opposed a moratorium on animal patenting and, to that end, vigorously contested the chief ethical and economic objections that had been raised against the practice.

Brill insisted that 'monster animals ... will not result from this technology'. Any organism's genes comprised a finely tuned ensemble. One could not 'take genes from a chicken and add them to a cow's chromosome and get a cow that lays eggs'. Godown dismissed as 'ridiculous' the Frankenstein scenarios that had been appearing in parts of the press – for example, the myth that 'genetic engineers are trying to develop some sort of humanoid slave containing the genes of humans and chimpanzees'. Wagner contended that genetically engineered animals posed no threat to the environment. Farmers were likely to keep them penned up and to retrieve any that did escape. Besides, animals, he explained, do not 'infect other animals [genetically] or humans except in sexual reproduction within their own species. A lot of us eat pork chops, a very few of us oink. [Pigs] cannot impart their genetic material to anything other than pigs' (US Congress, 1987b, pp. 38, 218, 259).

All three argued that genetic engineering, rather than increase animal suffering, would in fact reduce it. The outcomes of conventional breeding were unpredictable because they joined in one animal half of each parent's genes. The process, Brill pointed out, could yield progeny that were 'healthy or unhealthy, large or small, commercially interesting or not'. Genetic engineering, in contrast, involved adding just one or at the most several genes to an animal's native complement. It was thus 'more specific, more predictable and faster than breeding', Brill said, implying that it was on the whole more humane, too. Wagner noted that agricultural animals had been modified from the wild type by breeding over 10,000 years. 'Anyone who believes [that] the chickens in Frank Perdue's chicken houses are living in a natural ecosystem is simply unrealistic', he said, adding that genetic engineering could make farm animals resistant to the diseases that afflicted them in their synthetic environments (US Congress, 1987b, pp. 38, 218–219).

Wagner claimed that patented animals would help rather than hurt the small family farmer. The prime breeding stock of the chicken industry were hybrids, whose characteristics were trade secrets held by just a handful of breeding firms that operated in closed arrangements with an equally small number of producers. Wagner warned that if genetically engineered animals were not patented, similar secrecy would prevail among their developers; they would make 'exclusive arrangement with vertical integrators', and 'we will see a concentration of agriculture in the way chickens have been concentrated'. Godown added that genetically engineered animals would 'enable farmers to produce leaner beef ... at lower cost ... an advantage in anybody's language' (US Congress, 1987b, pp. 39, 259; see also Bugos, 1992).

Kastenmeier was convinced by much of the defence of animal patents. He expected that genetic engineering of animals would proceed with or without the encouragement of the patent system. A moratorium would only set a bad precedent and irritate the patent community without achieving any good purpose. Besides, he was convinced that patents were superior to trade secrets, and he thought that even opponents of animal patents would prefer openness to secrecy in agricultural research and development (Kastenmeier, 1987, 1991).

He also felt that while many of the ethical issues raised by the witnesses were legitimate, they did not fall within the realm of patent policy. In the USA, patent policy was literally amoral; it dealt only with the establishment and scope of property rights in innovations. It excluded others from making, using, or selling the property, but it did not by fundamental statute categorically exclude any invention from patentability.[2] Kastenmeier believed that issues such as whether the genetic engineering of animals might unduly foster animal suffering were within the regulatory powers of Congress but not within the jurisdiction of his subcommittee (Kastenmeier, 1987, 1991).

He was certain that the issue of patents on human beings did fall within it and he remained convinced that Congress needed to address it now rather than later. In keeping with his concern for both agriculture and the nascent biotechnology industry, he hoped to strike a balance between the creators and the users of animal patents. He repeatedly pressed witnesses for their views on a 'farmer's exemption' – whereby farmers could do what they wished with the offspring of patented animals, including selling them without paying a fee to the patent owner. Michael Ostrach, a senior official with the Cetus Corporation, a biotechnology company that had formed Agracetus, strongly objected. A farmer's exemption, he predicted, would only lead patent holders to charge farmers a higher price for the animal, thus making it very expensive (US Congress, 1987b, p. 296). In 1988, Kastenmeier nevertheless produced a bill that would exempt farmers from any restraint, including the restraint of royalty payments, on what they did with the progeny of their patented animals. It declared explicitly that human beings could not be patented. The bill passed the House, but it was not taken up in the Senate before the end of Congress (US Congress, 1988).

Kastenmeier attempted to advance another bill on the subject in 1990 but failed, and, defeated in the 1990 primary election, he was thereafter unable to push the issue. Since then, Congress has generally ignored the issue of animal patents. One reason is the force of the biotechnology complex in the political economy of intellectual property, but another is that the genetic engineering of animals has not fulfilled expectations, either in improving agricultural animals or in turning them into factories for molecular farming (*New York Times,* 2001). Only

[2] In contrast, Article 53 (a) of the European Patent Convention – which was established in 1962 and governs the national patent systems of its adhering nations – prohibits patents on any invention that is contrary to public order or morality. When Harvard University applied for a patent on its oncomouse, it had to demonstrate that the creation of Leder and Stewart did not violate the article (Kevles, 2002).

about 25 animal patents have been granted in the USA, most of them on laboratory mice and rats. However, if and when the genetic engineering of animals becomes scientifically achievable enough to figure in the political economy of agriculture, animal patenting may once again find a place on the public agenda.

Acknowledgement

I am grateful to the Andrew W. Mellon Foundation for support of research.

References

Atkinson, S. (1990) Interview with the author.

Brinster, R. and Palmiter, R.D. (1986) Introduction of genes into the germ line of animals. *The Harvey Lecture*, Series 80, pp. 1–4.

Bugbee, B.W. (1967) *Genesis of American Patent and Copyright Law*. Public Affairs Press, Washington, DC, 152.

Bugos, G.E. (1992) Intellectual property protection in the American Chicken-breeding industry. *Business History Review* 66, 127–168.

Bugos, G.E. and Kevles, D.J. (1992) Plants as intellectual property: American law, policy, and practice in world context. *Osiris*, 2nd Series, VII, 119–148.

Diamond vs. Chakrabarty (1980), 447 U.S. 303, 100 S. Ct. 2204, pp. 2206–2212.

Ex Parte Allen (1987) *United States Patent Quarterly*, 1425.

Gordon, J.W., Scangos, G.A., Plotkin, D.J., Barbosa, J.A. and Ruddle, F.H. (1980) Genetic transformation of mouse embryos by microinjection of purified DNA. *Proceedings of the National Academy of Sciences USA* 77, 7380–7384.

Jones, R.A. (1987) 'Biotechnology, in search of a more perfect pig.' *Los Angeles Times*, 12 July, 1.

Kevles, D.J. (1986) Unholy Alliance. *The Sciences*, September/October 25–30.

Kevles, D.J. (1994) Ananda Chakrabarty wins a patent: biotechnology, law, and society, 1972–1980. *HSPS: Historical Studies in the Physical and Biological Sciences* 25, 111–136.

Kevles, D.J. (2002) Of mice and money: the story of the world's first animal patent. *Daedalus* 78–88.

Machlup, F. (1968) Patents. In: Sills, D.L. (ed.) *International Encyclopedia of the Social Sciences*. Macmillan, New York, XI, pp. 461–464.

New York Times (2001) Biotechnology venture hits unexpected snags. 23 November, C5.

Palmiter, R.D. (1990) Interview with the author.

Palmiter, R.D., Brinster, R.L., Hammer, R.E., Trumbaur, M.E., Rosenfeld, M.G., Bimberg, N.C. and Evans, R.M. (1982) Dramatic growth of mice that develop from eggs microinjected with metallothionein-growth hormone fusion genes. *Nature* 300, 611–615.

Palmiter, R.D., Norstedt, G., Gelinas, R.E., Hammer, R.E. and Brinster, R.L. (1983) Metallothionein-human GH fusion genes stimulate growth of mice. *Science* 222, 809–814.

People's Business Commission (PBC) (1979) *Brief on Behalf of the People's Business Commission, Amicus Curiae*. In: *Diamond vs. Chakrabarty*, U.S. Supreme Court, Docket No. 79–136, 447 U.S. 303, January 1980.

Pursel, V.G., Miller, K.F., Bolt, D.J., Pinkert, C.A., Hammer, R.E., Palmiter, R.D. and Brinster, R.L. (1989a) Insertion of growth hormone into pig embryos. In: Heap, R.B., Prosser, C.G. and Lamming, G.E. (eds) *Biotechnology in Growth Regulation*. Butterworth, London, 1818–1886.

Pursel, V.G., Pinkert, C.A., Miller, K.F., Bolt, D.J., Campbell, R.D., Palmiter, R.D., Brinster, R.L. and Hammer, R.E. (1989b) Genetic engineering of livestock. *Science* 244, 1281–1288.

Rensberger, B. (1984) Scientists hail gene transfers as promising. *Washington Post*, 20 November, 1.

Russell, C. (1984) USDA using human gene in effort to grow super livestock. *Washington Post*, 1 October, 1.

Saltus, R. (1988) Era of 'designer animals' looms. *Boston Globe*, 26 December, 63.

Toronto Star (1988) Genetic tests open scary world of super species. 29 March, A27.

US Congress (1987a) House, Hearings before the Subcommittee on Courts, Civil Liberties, and the Administration of Justice, Committee on the Judiciary. *Patents and the Constitution: Transgenic Animals*. 100th Congress, 1st Session, 11 June, 22 July, 21 August and 5 November, 1987, Reporter's Minutes, in possession of the Committee.

US Congress (1987b) House, Hearings before the Subcommittee on Courts, Civil Liberties, and the Administration of Justice, Committee on the Judiciary. *Patents and the Constitution: Transgenic Animals*. 100th Congress, 1st Session, 11 June, 22 July, 21 August and 5 November, 1987.

US Congress (1988) H.R. 4970, in House, *Congressional Record*, 13 September, 1988, H7436-H7438.

US Congress (1989) House, Hearings before the Subcommittee on Courts, Civil Liberties, and the Administration of Justice, Committee on the Judiciary. *Transgenic Animal Patent Reform Act of 1989*, 100th Congress, 1st Session, 13 and 14 September, 1989.

The Economics of Patents

3

Corinne Langinier and GianCarlo Moschini

Department of Economics, Iowa State University, Ames, IA 50011, USA

Introduction

Patents are perhaps the most important legal instruments for protecting intellectual property rights (IPRs). A patent confers to an inventor the sole right to exclude others from economically exploiting the innovation for a limited time (20 years from the date of filing). To be patentable, an innovation must be *novel* in the sense of not constituting part of the prior art or more generally of not being already in the public domain. A patentable innovation also must involve an inventive step, meaning that it must be *non-obvious* to a person with ordinary skills in the particular field of application. The innovation also must be *useful* to be patentable; that is, it must permit the solution of a particular problem in at least one application. A major element of a patent application is *disclosure*: the invention must be described in sufficient detail to *enable* those skilled in the particular field to practise it. The patent application also lays out specific claims as to the scope of the patent itself. The traditional statutory scope of patents, encompassing machines, industrial processes, composition of matter and articles of manufacture, excluded important kinds of scientific discoveries such as laws of nature, natural phenomena and abstract ideas. However, recent developments in the use of patents for computer software, information technology and biotechnology innovations are challenging a reductive interpretation of such exclusions. See Merges (1997) for further details and a comprehensive treatment of US patent law.

Given that private property is inherently associated with the freedom of choice that constitutes the cornerstone of market economies (Barzel, 1989), it may seem obvious that intangible assets associated with human inventiveness and creativity should enjoy a legal status similar to that afforded to the ownership of other more standard goods and services. However appealing, this consideration does not quite do justice to the many aspects of the economics of patents. There are at least two relevant approaches one can take to articulate a fuller discussion of this intellectual property (IP) institution, a positive analysis and a normative one. From a positive perspective, we may ask how patents, as currently implemented, affect the workings of the economic system. Ultimately, we wish to understand how the existence of patents affects the allocation of resources to, and the distribution of income arising from, inventive activities. A positive economic framework also can be used to ask the related question of why patents came to exist. Alternatively, from a more normative perspective, we can explore whether the institution of patents is a desirable attribute of the economic system, and what the features of an optimal patent system are. In what follows, we consider these economic questions in more detail, focusing on conceptual and theoretical analyses.[1]

Benefits and Costs of Patents

From an economic perspective, the crucial features of patents are that: (i) they deal with new knowledge, as embodied in an innovative product or process, and (ii) they confer (limited) monopoly rights to the inventor. New knowledge that makes possible the production of new products and/or processes obviously carries considerable economic value, but it has features that make it problematic for the market system to handle properly (Arrow, 1962). Specifically, knowledge is a quintessential public good. Pure public goods have two basic attributes. First, they are nonrival in consumption, meaning that a person's use of a public good does not affect the amount of it that is available for others. Second, they are non-excludable, meaning that it is not possible to prevent individuals from enjoying the public good once it is available. An example of a pure public good is national defence. It is clear that, without IPRs, most discoveries and inventions would exhibit public good attributes.

The problems that a competitive system has with public goods are readily apparent. An inventor may bear all the cost of an innovation, but everyone benefits (possibly to varying degrees) from a discovery, and thus everyone has an incentive to free ride on the innovative efforts of others. The inherent externalities associated with this class of public goods generate a market failure: a competitive market system may be

[1] Although this body of work reflects accepted stylized facts and empirical findings, as well as the evolving legal and institutional features of the patent system, there is also a distinct and large set of empirical studies on the economics of patents that we will not review here.

expected to provide an inefficiently low level of innovations. IPRs in general, and patents in particular, address this problem by attacking the non-appropriability of knowledge that lies at the heart of this market failure. Specifically, by endowing innovators with property rights on their discoveries, patents are a legal means of affecting the excludability attributes of an otherwise pure public good.

The main economic benefits and costs of the patent system are intimately related to nature of the market failure that it addresses, and to the second-best character of the solution that it provides. We begin the discussion of such effects along the taxonomy suggested by Mazzoleni and Nelson (1998).

Patents can promote new discoveries

By endowing discoverers with property rights over the fruits of their efforts, patents affect the incentive to innovate and are likely to increase the flow of innovations. This increase is presumably desirable, given that otherwise the market system may provide too little new knowledge, but by giving the patentee exclusive rights on the exploitation of a unique economic good that is still non-rival in consumption, a patent creates a monopoly situation that adversely affects the efficient use of new knowledge.

Nordhaus (1969) and Scherer (1972) provided an early formalization of this fundamental trade-off between benefits and costs of the patent system. For a simpler illustration, consider the potential market for a new product, say, a new pharmaceutical. If p denotes the price charged to consumers, the latent demand $D(p)$ is likely to display a choke price $\bar{p}$ such that $D(p)=0$ for all $p \geq \bar{p}$ (there exists a prohibitive price that drives demand to zero). Furthermore, this latent demand is downward sloping, i.e. $D'(p)<0$ for all $0 \leq p < \bar{p}$ (a lower price makes the drug attractive to an increasing number of potential users). Suppose that it takes a total cost F to develop this drug, including the testing required for approval, but once this knowledge is available it takes only a unit cost c to actually manufacture and sell the drug. This situation is represented in Fig. 3.1, where demand applies for a specific period (a year, say). Given some qualifications, area $(S+\Pi+L)$ represents the per-period 'social surplus', i.e. the monetary benefits that would accrue to society if this new product were developed (and produced at the efficient level q^C). If the present discounted value of the current and future stream of such benefits exceeds the research and development (R&D) cost F, then net benefits are possible for society from the introduction of the new drug. However, without IPRs, no individual has an incentive to incur the cost F: the ability of competitors to readily copy the innovation would drive the market price to its unit production costs c, and the cost F cannot be recouped.

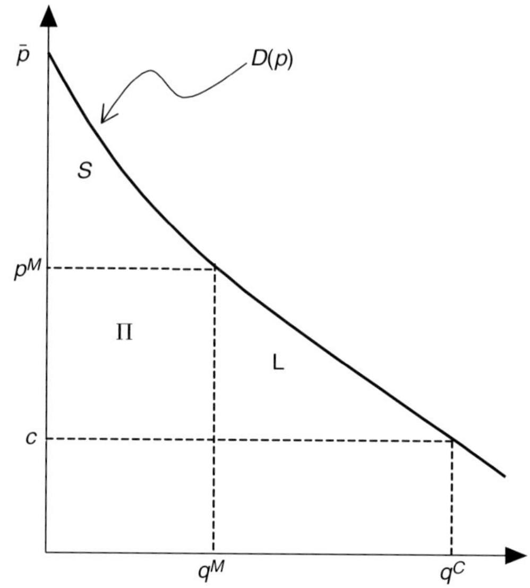

Fig. 3.1. Patents and the market for a new product. See text for explanation.

However, with a patent, the innovator becomes a monopolist in the market for the new drug and can profit by pricing the drug optimally at p^M, where $c < p^M < \bar{p}$. The innovator's per-period profits are represented by the area Π in Fig. 3.1. If the present discounted value of such profits, over the life of the patent, exceeds the R&D cost F, then a sufficient incentive exists for this innovation to be brought about, and society as a whole enjoys the benefits $(S + \Pi)$ for the duration of the patent, and benefits $(S + \Pi + L)$ thereafter. But note that, during the life of the patent, the innovation is produced at an inefficiently low level from a social point of view (i.e. $q^M < q^C$).[2] This brings to the fore a fundamental trade-off of the patent system: the balancing between the benefits of encouraging additional innovative activities and the costs of foregoing the competitive provision of some goods and services. *Ex post*, that is, given that an innovation is available, a monopoly is bad from society's point of view because it restricts uses of the new product and/or process (relative to the competitive provision of the innovation). Earlier economic analysis focused extensively on this issue and questioned the economic desirability of the patent system (Machlup and Penrose, 1950), but the profit opportunity created by the monopolistic control of the innovation can be a powerful *ex ante* incentive, presumably enough to motivate R&D investments that would otherwise be neglected.

---

[2] Here, S measures the consumer surplus while the patent is active, and $(S+\Pi+L)$ is the consumer surplus after the patent has expired. Thus, L is the 'loss' to society from under-utilizing the innovation during the life of the patent.

Patents can help the dissemination of knowledge

An additional benefit of patents is related to the disclosure requirement. In most countries, patents are disclosed 18 months after the filing date or earlier. The importance of this feature is predicated on the fact that, without patents, inventors can rely on trade secrets to protect their discoveries (Friedman *et al.*, 1991). By providing an incentive for disclosure, patents are held to contribute to a desirable dissemination of scientific and technical information, allowing other inventors to avoid duplicating existing discoveries and making it easier to develop further innovations that build on the known state of the arts (possibly by 'inventing around' a patent as well). It is useful to note that the disclosure argument offers an economic role for patents, even for inventions that have already taken place, and as such it is quite distinct from the incentive role of patents due to the increased appropriability of R&D output.

Patents can avoid wasteful innovation efforts

As the arguments on the disclosure property of patents suggest, an important and beneficial effect of new knowledge is that it makes possible further innovations and discoveries. Discoveries from basic research are often of this sort, at times opening up entire new fields of research. It can be argued that patenting of such seminal inventions can have useful social pay-offs. This rationale is articulated in the so-called prospect theory of patents, originated by Kitch (1977). It relies on the notion that broad, early property rights on key inventions allow an orderly pursuit of follow-up innovations and reduce wasteful innovation races. An analogy can be made to the practice of granting mineral claims on land where no discovery has yet been made, to avoid a wasteful mining of the prospect. Whereas patents in such cases can clearly have positive efficiency effects, it is also easy to see that broad, early patents can adversely affect further research, especially when the original discovery has applicability in many uses. If the original inventor does not have a comparative research advantage or interest in pursuing some research directions, and licensing of the patented innovation to third parties is problematic (perhaps because of excessive transaction costs), patenting can have adverse effects on the flow of further innovations. We will return to related issues later when discussing the appropriate breadth of patents.

Patents can help technology transfer and commercialization

Whereas the difficulty of licensing may reduce the desirability of exceedingly broad property rights, patents can actually play a critical role in licensing and, more generally, in the dissemination of new knowledge. A key insight here is due to Arrow (1962), who stressed the information nature of new knowledge and noted a peculiar property in the determination of demand for information. Specifically, to put a value on information a would-be purchaser needs to have the information, but at that point, if the seller does not have property rights on the information, the would-be buyer has no more incentive to pay for it. Patents, therefore, can play a crucial role in reducing transaction costs of licensing innovation and, more generally, in technology transfer.

A related but distinct role for patents as instruments of technology transfer has been articulated to rationalize the Patent and Trademark Laws Amendments of 1980, commonly known as the Bayh–Dole Act. The main elements of this reform were to allow universities and other entities to patent, retain title to and commercialize federally funded inventions, and to allow federal agencies to grant exclusive licensing for their inventions. Based on the view that the main role of patents is to provide incentives for innovation that would not occur otherwise, it would be difficult to make an economic case for public institutions patenting discoveries that already have been publicly funded and accomplished. Likewise, the role of patents in transferring information would be irrelevant in this case, because public research institutions have little use for trade secrets, and because it is difficult to improve on the dissemination of information achieved by simply publishing a discovery. However, the presumption here is that many discoveries produced by publicly funded R&D, and in the public domain, may not be used in technological developments because, without an exclusive licence backed up by patent rights, firms would not be interested in expensive development work required to transform an invention into a new product. Little evidence exists to corroborate this belief for the case of university patenting (Mowery *et al.*, 2001), but more generally, when it is difficult to assert property rights on development work, a patent on underlying innovations obviously may aid development and commercialization of new technologies.

Length and Scope of Patent Protection

As illustrated in the foregoing, the patent system entails a fundamental trade-off: inventors are given a monopoly position (which entails inefficiencies) in order to provide them with incentives to innovate (which carries economic benefits).[3] In other words, the dynamic efficiency of

[3] Here and in what follows, we do not distinguish between the incentive to create and the incentive to commercialize an innovation.

encouraging invention leads to static inefficiency created by a monopolistic situation. Given that, a natural question to ask is, what is the optimal degree of patent protection? The degree of market power provided by a patent essentially depends on three elements: the length (duration) of the patent, the breadth of the patent and the height of the patent. We analyse these features of patent protection for the case of a single innovation first, and consider the issues related to cumulative innovations later on.

Duration of the monopoly power

The length of the patent determines the duration of monopoly power that a patentee can expect and thus affects the static inefficiency as well as the incentive role of patents. The benefit to the innovator increases with the length of the patent, but from society's point of view, too short a patent may dissuade research, whereas a patent that is too long may give excessive rent to the owner and may block further improvement.

Consider first a given innovation, such as the new product case illustrated in Fig. 3.1, where Π denotes the per-period profit accruing to the patentee. Here, the optimal duration of the patent should be determined so as to recoup exactly the R&D expenditure necessary to bring about the innovation. Thus, for a discount rate r, and given a one-time research cost F of bringing about the innovation, the optimal patent length T^* would solve $\int_0^{T^*} \Pi e^{-rt} dt = F$. A more realistic model would allow the size of the innovation to vary with the investment in R&D, which in turn depends on the length of the patent. For simplicity, think of a process innovation that leads to a reduction in the unit cost of producing a good. The optimal patent length must now balance the social gain from the innovation due to a larger reduction in cost and the social loss associated with a longer delay in the exploitation of the innovation by rival firms. This point was originally made by Nordhaus (1969), who found that an optimal patent must be of finite duration but strictly positive, as shown later by Scherer (1972).

In these earlier models, it was assumed that R&D investment always leads to an innovation the size of which depends on the investment undertaken and where an R&D project necessarily succeeds. This is a restrictive assumption that does not take into account competition at the R&D level. The introduction of such competition can, under certain circumstances, increase the optimal length. Indeed, the expected pay-off of each competitor is reduced; therefore, for a given number of competitors, an increase in the patent duration may be the only way to re-establish the R&D incentive (Kamien and Schwartz, 1974). However, in other circumstances, competition can reduce the required patent length, as is the case in DeBrock's (1985) model, where it is assumed that competition takes place at the research level and not at the development level where patent holders can develop their innovation without threat of competition. Here, R&D investment at the development stage is the same as without competition, but competition at the research level leads to useless duplication of costs for firms that cannot get a patent, and that in turn reduces the social surplus.

The industrial context may play an important role in the determination of the optimal patent length. For instance, in markets where demand is very elastic (i.e. quantity demanded is very responsive to a small change in price), monopolistic pricing of the innovation can lead to large *ex post* welfare losses (i.e. a large L in Fig. 3.1). Hence, in such markets a shorter patent length would be warranted. The optimal patent length is also likely to be innovation-specific, and thus the one-size-fits-all patent length (20 years from filing) is sub-optimal. Roughly speaking, a uniform patent length provides too much protection for 'easy' innovations (those that would have been pursued even with a shorter patent period), thereby creating unnecessary efficiency losses, whereas it provides too little protection for 'difficult' innovations, such that some research projects that are socially desirable do not get undertaken.[4]

The scope of a patent

Whereas the length of the patent protection characterizes the duration of the monopoly power, the scope of a patent bears on the intensity of the induced monopoly power (Merges and Nelson, 1990). The breadth of a patent defines the range of products that are encompassed by the claims of the patent and therefore protects the patent holder against potential imitators. In general, the less specific the claims of the patent are, the broader the patent. The height of a patent, on the other hand, confers protection against improvements or applications that are easy or trivial. The value of a patent to a firm depends on how effective its protection is in these two dimensions (breadth and height), in addition to being related monotonically to the patent length.

Unlike its maximum length, which is fixed by law, patent breadth is, to a certain extent, endogenous. Patent breadth depends on the claims put forth by the patentee and also is a feature over which the patent

[4] These considerations are somewhat tempered by the fact that effectively different patent lengths can be implemented through the imposition of different renewal fees (Cornelli and Schankerman, 1999). In most European countries, patent holders must pay an annual renewal fee in order to keep their patents in force. In the USA, only three renewal fees have to be paid: at patent ages 3, 7 and 11.

office has some discretion (at the examination stage).[5] Clearly, a reduction in the breadth of patents would induce more competition (e.g. imitation), which benefits consumers, but too narrow a patent reduces the incentive to innovate. What is the optimal breadth for patents? Economic analysis suggests two kinds of results in this setting. Narrow and long patents can be found to be optimal because broad patents are costly for society in that they give excessive monopoly power to the patent holder (Gilbert and Shapiro, 1990). Central to this conclusion is that the flow of pay-off from holding a patent has a negative impact on the social surplus. Thus, a minimum level of flow of pay-off (breadth), with duration adjusted accordingly, would be socially optimal. In Klemperer's (1990) more general model, both narrow and long, or broad and short, patents can be optimal, depending on the structure of demand. Broad and short patents can also be optimal when they discourage imitation and thus enhance the incentive to innovate (Gallini, 1992). Specifically, imitation is discouraged when it is too costly (broad patent) and when imitators do not have enough time to enter the market (short patent).

Denicolò (1996) analyses these alternative conditions in a single model, in which the problem is to minimize a ratio of social loss to the incentive to innovate. The breadth influences both elements of the ratio, and, depending on the assumptions about the nature of competition in these markets, all the above results can be found. Thus, in different industries, different optimal patent policy could be enforced. Patents in biotechnology should be different than patents in e-commerce, for example. Note that in these models, patent length and the scope of protection are substitutes for the purpose of providing a given level of protection to the innovator. This is not necessarily the case when we consider cumulative innovations, as we will see below.

Another dimension of the patent right is the protection it gives to the innovator against improvements that are too close to the patented innovation. This feature is sometimes referred to as the novelty requirement, also called 'height' (van Dijk, 1996) or 'leading breadth' (O'Donoghue *et al.*, 1998). Competition from follow-up innovations are clearly affected by the height of the patent, and too high a patent gives excessive monopoly power to the patent holder. As for the interaction between height and duration, La Manna (1992) shows that a patent of infinite duration and finite height can be optimal.

In conclusion, patent scope (breadth and height) can substitute for patent length in providing *ex post* returns from (and therefore *ex ante* incentive for) an innovation, but one should not conclude that broader

[5] In US case law, breadth and height of a patent also are upheld by the so-called doctrine of equivalents, which asserts that a new product that is outside the stated claims of an existing patent could still infringe on this patent if it is essentially an equivalent product.

or higher patents necessarily induce more research. The problem is that, *ex ante*, a potential inventor needs to consider the possibility of discovering something that may infringe on existing patents. The broader or higher the patents, the greater the risk of such a possible unwanted infringement, and this possibility can discourage R&D investments. As discussed in the foregoing, specific conclusions about optimal patents are sensitive to the way one chooses to model the innovation process. Also, the degree of substitutability between length, breadth and height of a patent depends on the nature of the innovation. Currently, there seems to be widespread concern about exceedingly broad patents being granted, particularly in recent innovation fields such as biotechnology and so-called business methods (Merges, 1999). This is particularly the case for cumulative innovations, when an innovation is useful mostly as a research tool in other R&D activities.

Cumulative innovations

The importance of patent height is most apparent for the case of sequential innovations. Such innovations are particularly relevant when the innovation is a research tool, used mainly in the R&D process for further innovations. Consider first the case of a basic innovation followed by its improvement, which eventually can be accomplished by the same innovator. What is the impact of the scope of the protection in the diffusion of innovations over time? Scotchmer and Green (1990) focus on the profit incentive for R&D when the second innovation is more profitable than the first one (from social and private viewpoints) and when there exists an information externality among innovators (patent disclosure confers a positive externality on the innovator's competitors). The policy question here is to determine how stringent the novelty requirement should be in order to protect the profit of innovators, while encouraging disclosure of innovation. It is presumed that the reasons for granting a patent are to create an incentive to do research and to accelerate aggregate innovation through disclosure of innovations. The first reason calls for a strong novelty requirement (any small improvement should be found to infringe upon the patented innovation and thus protect fully the first innovator), whereas the second reason calls for a weak novelty requirement.

Because the information externality that the innovator may want to avoid is due to the disclosure requirement of patents, the innovator may prefer to keep small innovations secret. The strategic suppression of small improvements can avoid the *ex ante* erosion of profit due to a weak novelty requirement (Scotchmer and Green, 1990). The weak novelty requirement seems to be socially preferable because an early disclosure accelerates the introduction of the improvement. Nevertheless, even in this system, the first innovation is not always patented. When the *ex ante* profit is larger with the strong novelty requirement, this last system is socially preferable. In any case, the first innovator can choose not to patent the innovation. One way to force firms to disclose their small innovations without undermining their profit would be to force them to cooperate. This could take the form of licensing (Green and Scotchmer, 1995; Scotchmer, 1991). In the same vein, Matutes *et al.*, (1996) show that a system based on an optimal height rather than an optimal length is socially preferable because it induces an earlier introduction of the innovation.

However, it is not always the case that the first innovator is able to improve on her innovation or even to develop applications. For instance, small laboratories that do not have access to sufficient capital may be unable to perform further development. In this setting, what is the appropriate patent scope that gives enough incentive to the first and second innovators to undertake R&D investment? Let us first assume that only one firm is capable of being the second innovator. In this setting, each innovator should receive an incentive at the level of the social benefit he or she creates. For the second innovator, it should be the incremental benefit created by his/her improvement, whereas for the first innovator, it should be the sum of the first innovation benefit and the incremental benefit. Indeed, the feasibility of the second innovation is due to the very existence of the first one. For this reason, the actual profit-sharing arrangement between innovators is important. If profit sharing is in favour of the first innovator, it will promote radical innovations, whereas if it is in favour of the second innovator, it will promote improvements.

It is nevertheless very difficult to favour improvement if the first innovation has no value by itself (but is fundamental to the discovery of the second innovation). On the other hand, a policy in favour of the first innovator will reduce follow-up improvements and will thus affect the profit of the first innovator. Public authorities have two ways to intervene in the profit-sharing process: through the scope of the patent (height, length or patentability of the second generation of innovations), and through the competitive policy in regard to licenses. Green and Scotchmer (1995) present a situation in which it is profitable for innovators to sign *ex ante* licence agreements. They show that an infinite height (i.e. any second-generation innovator needs to get a licence from the original innovator) is optimal when there is certainty about the return of the second generation of innovations. In the case of uncertainty, that is no longer the case.

Scotchmer (1991) explores the role of the patentability of the second generation of innovations when the improvement of the initial innovation can be done by more than one firm. The very existence of competition to obtain the second patent reduces the expected profit of the first innovator and thus reduces the attractiveness of the *ex ante* licence. An *ex ante* licence agreement can be impossible to implement in situations where the second generation of innovators are reluctant to disclose their ideas for improvement to the first innovator. Chang (1995) considers this situation and shows that, no matter what the value of the first innovation, there exists a threshold value for the second innovation below which a licence is required and above which it is not required. The smaller the benefit of the first innovation, the higher must be the patent in order to provide enough incentive to the first innovator, but optimal patent height is not a monotonic function of the social value of the first innovation, and when this value is large, the patent also must be high.

To capture fully the impact of the cumulative nature of the innovations on the incentive to innovate, it is useful to consider a dynamic model where improvements of the innovation arise randomly (O'Donoghue *et al.*, 1998). In this setting, the innovators must be protected against improvements (leading breadth or height) and against imitations (lagging breadth or breadth). Here it is assumed that the probability that the same innovator makes two successive improvements is almost nil. Depending on the rate at which improvement ideas arise, under a perfect lagging breadth, innovators may over-invest or under-invest. They will tend to under-invest if ideas are too frequent because they fear that another improvement will be introduced too early, but if ideas are not that frequent, firms will over-invest because each innovator expects to increase her pay-off once she becomes a follower after having been a leader. With a maximal lagging breadth, and if leading breadth is finite but duration is infinite, firms under-invest in R&D. An infinite duration for a low patent reduces the cost associated with the delay in diffusion (O'Donoghue *et al.*, 1998).

Thus, when innovation is cumulative, the acceleration of the disclosure of the first innovation is crucial to allow further improvements, but this must be done without altering the incentives of the first innovator. Furthermore, when innovations are cumulative, their complementarity gives rise to additional issues.

Complementary innovations

To manufacture a complex new product (e.g. a computer chip or a transgenic crop), it is common for firms to need a number of intermediate inputs that are patented, with such patent rights likely held by different firms. Insofar as the patented inputs are highly complementary, or even essential, in the manufacture of the new product, the suppliers of these inputs essentially hold blocking patents. The danger then is that the manufacturer becomes susceptible to *hold-up* by the patentees (Shapiro, 2001). In the context of biotechnology innovations, this problem has been characterized as the 'tragedy of the anticommons' (Heller and Eisenberg, 1998). Recall that the standard notion of tragedy of the commons refers to the over-use of a common property resource, and it arises because not enough property rights are allocated. The tragedy of the anticommons situation arises because, somewhat paradoxically, there is an excess allocation of property rights (i.e. too many gatekeepers with the right to levy a tax), and this situation can lead to an under-use of the resource (in this case, the pool of knowledge).

The hold-up problem when patented innovations are complements can be quite damaging to the functioning of the patent system. The issue here is mostly one of coordination among patent holders. Mechanisms that can address the hold-up problem in this setting include cross-licensing and patent pools. With cross-licensing, which can be implemented with or without licence fees, two (or more) firms agree to share a subset of each other's patented technologies. This kind of arrangement is particularly attractive when the relevant patents are held by the manufacturers of the new product that needs the complementary patented inputs. Cross-licensing is apparently quite common between firms engaged in the design and manufacture of microprocessors, for example. A patent pool is an arrangement whereby a set of patents is licensed in a single package, for a posted fee, either by a single owner or by an entity especially set up to handle this arrangement. The obvious danger here is that, whereas the inclusion of complementary patents in the pool is justifiable (as detailed in the foregoing discussion) and leads to pro-competitive behaviour, the inclusion of substitute (rival) patents in the pool could be a means to implement collusion and could exacerbate non-competitive behaviour. Patent pools, therefore, can pose anti-trust issues. Another way to reduce the fragmentation of patent ownership is through mergers and acquisitions, and that, indeed, appears to be one of the driving forces behind the recent industry consolidation in the life sciences sector.

Patents, Market Structure and the Incentive for R&D

The research activity of innovators undoubtedly is influenced by the patent system, as well as by the industry structure. To gain some insights into the interaction of patents and industry structure, we explore how innovators invest in R&D. Note first that the discovery of an innovation affects the structure of the market. This is easy to understand if we consider a process innovation that allows reducing production costs: firms can compete in the pre-innovation market and, if a patent is granted, the holder will be a monopolist in the post-innovation market. Competition can also take place at the R&D level. The notion that firms may be competing at the innovation stage has been a recurring one in this chapter. This competition is very much affected by the patent system, which has the feature of a winner-takes-all contest. To analyse the effects that this may have on R&D, recall that the presumption is that, without patents, economic agents do not have sufficient incentives to invest in R&D activities and, consequently, not enough innovation activity would take place. The term of comparison here is usually the 'socially optimal' level of R&D activities, which one can define (at least conceptually) taking into account the overall expected benefits brought about by an innovation, as well as the costs of R&D projects. This first-best solution typically cannot be achieved by a decentralized market system, for lack of incentives and coordination mechanisms (the 'market failure' predicament), and it cannot be achieved by a centralized government, either for lack of information and implementation mechanisms. Given these benchmarks, the patent system is likely to increase private investments towards the first-best (socially optimal) level of R&D.

To isolate the pure incentive to innovate (that is, without any competition at the innovation stage), it is necessary to note that the value of an innovation to an inventor depends on the nature of the innovation but also on the market structure that applies at the production stage (Arrow, 1962). Monopoly and competition provide useful polar cases. It turns out that, *ceteris paribus*, the incentive to innovate is lower in a monopolistic market than in a competitive market. This is due to the replacement effect: a monopolist has less to gain from a given innovation (say, a cost-reducing innovation) because one monopolistic situation replaces another monopolistic situation, but even under competitive production conditions, the incentive to innovate is less than what is socially optimal. This latter case is actually depicted in Fig. 3.1, which can be interpreted as representing a market where competitive producers with unit cost c must pay a royalty to the inventor. In this situation it is optimal for the inventor to set the royalty rate at $(p^M - c)$, and thus the incentive to innovate is the profit Π. But, from society's perspective, the incentive to innovate ought to be the entire (potential) consumer surplus $(S + \Pi + L)$.

What is missing from the foregoing analysis is the possibility of entry in the innovation industry; typically it is possible (and likely) for more than one firm to pursue the same research result. The resulting 'patent race' tends to dissipate the expected rent from patenting and to increase aggregate R&D investment (in a fashion akin to the 'common pool' problem of competitive fishing in an open access fishery). To see what some of the implications might be, following Wright (1983) and Tandon (1983), suppose that n firms pursue the same invention. Each firm incurs a one-time research cost F. Let $P(n)$ represent the probability that the invention takes place (at least one firm is successful). In general, more firms increase the aggregate success probability but at a decreasing rate (i.e. $P'(n) > 0$ and $P''(n) < 0$). For simplicity, assume that the social benefits from the discovery are equal to the private benefits that accrue to the inventor, say B. From a social point of view, the optimal number of firms n^* that pursue the invention should maximize the expected net social benefit and thus should solve $P'(n^*)B = F$ (i.e. expected marginal social benefit = marginal social cost). However, because each individual firm has an incentive to join the patent race as long as the expected profit is greater than the cost F (under risk neutrality), the equilibrium number of firms n^C in competition solves $P(n^C)B/n^C = F$. Given the properties of the aggregate probability function $P(n)$, it then must be that $n^C > n^*$, and therefore excessive resources are allocated to R&D by this patent race.

Consider now the dimension of time and suppose that the date of discovery of the invention is a decreasing function with the investment in research (Dasgupta and Stiglitz, 1980), and assume that the investment is completely paid at date zero. The pre-innovation situation is competitive, and only the firm that first makes the innovation will be granted a patent (monopolistic post-innovation situation). This situation is similar to an auction in which at date zero the patent will be granted to the firm that proposes to innovate first. In this setting, only one firm will invest exactly the amount that it will gain from innovating. The deterministic nature of the relationship between the date of discovery and the investment makes it easy to find that the investment undertaken by a monopolist is smaller than the investment undertaken by competitive firms. Furthermore, the monopolist invests less than that which is socially optimal. This is due to the negative externality of the investment of an innovator on her rivals, which none of the firms internalize.

In this setting, what happens to the incentives to innovate if a potential entrant can introduce the innovation instead of the incumbent? In other words, do technological innovations come from a monopolist in the pre-innovation market or from new innovators? The pure incentive to invest is higher for potential entrants, due to the replacement effect, but there is another effect that has to be considered: the efficiency effect that arises because the profit of a monopoly is bigger than the sum of the duopoly profits. Thus, a monopolist has more to lose when a competitor enters the post-innovation market compared to a potential entrant. This leads to preemption (Gilbert and Newbery, 1982) and to the persistence of monopoly.

Now we add one more degree of realism and consider not only that discovery depends on the R&D investment but also that it is random, as in the case of 'patent race' models with stochastic innovation (Loury, 1979; Lee and Wilde, 1980; Reinganum, 1989). The date of discovery is no longer deterministic, but random according to a Poisson process. The investment can be fixed and incurred at the beginning (Loury, 1979), or it can be incurred as a constant flow over time (Lee and Wilde, 1980). The advantage of the second model is that firms can stop to invest as soon as a rival has made a discovery. These models have no memory: the probability of success does not vary as time goes by or as investment is accumulated. Each firm considers only its own marginal benefit and does not take into account the reduction that its own investment will impose on its rival's payoff. It is not known when a discovery will be made (technological uncertainty), nor who will make it (market uncertainty). Loury (1979) shows that as the number of firms that compete increases, the average optimal investment decreases. On the contrary, in the model of Lee and Wilde (1980), it increases. The difference in results is due to the different structure of investment. Nevertheless, for a given number of firms engaged in the race, firms tend to over-invest in R&D compared to the socially optimal level in both models. Thus, there is a tendency for excessive research.

In this setting, do monopolists have a stronger incentive to invest than do potential entrants? Which effect (replacement or efficiency) dominates the other depends on the nature of the innovation. For an innovation that is almost drastic, a monopolist invests less than a potential entrant and thus preemption disappears (Reinganum, 1983).[6] There is no efficiency effect anymore; only the replacement effect plays a role. When the firm behind in the race can leap-frog, or when the process is without memory (i.e. at each point in time the probabilities of winning are the same for each firm), there is no pre-emption (Fudenberg *et al.*, 1983). On the other hand, when it is not possible to leap-frog, as is the case in Gilbert and Newbery (1982), then there is always persistence of monopoly.

Thus, firms engaged in a patent race may collectively invest more than that which is socially desirable. Furthermore, firms engaged in a patent race are likely to choose research strategies that are too risky. Because the race is essentially a winner-takes-all contest, there is a clear pay-off to being first, but the penalty for losing does not depend on how far behind a firm ends up, and firms will tend to choose research projects that, from a social perspective, display too much correlation (i.e. there is too much duplication of research efforts; Dasgupta and Maskin, 1987). To illustrate this last point, suppose that it is socially desirable for a particular firm to choose a research direction very different from that of its competitors (because that would increase the probability of someone being successful). The reason an individual firm may not want to follow such a (socially optimal) strategy is because by doing so it would increase the probability that a competitor is successful in instances when the firm is not.

[6] As in Arrow (1962), an innovation is 'drastic' if its pricing is not affected by the threat of competition (the innovator can behave as an unrestricted monopolist).

Beyond Patents

The main justification for patents is to foster innovation in a market economy, but the patent system is not the only method for encouraging innovation. Copyrights and trademarks, of course, are additional instruments for IPR protection that typically apply in contexts where patents do not. Trade secrets, on the other hand, can apply to patentable innovations and can provide effective protection against another party's discovery by inappropriate means (although a trade secret offers no protection against independent discovery or reverse engineering). Certain biological innovations are also afforded *sui generis* protection by such means as plant patents and plant variety protection certificates (Barton, 1998).

More generally, alternatives to patents include rewards or prizes, procurement contracts and public production of new knowledge. With the reward system, the government specifies a fixed sum of money for a well-defined research goal and then awards this 'prize' to the first firm to achieve the desired result. Asymmetric information between researchers and the government can make it difficult to implement the reward mechanism (Wright, 1983). Specifically, to be effective, the government must know about the feasibility of various research projects as well as be able to assess the demand for various potential innovations. But firms are likely to be better informed than the government on such matters, and a decentralized solution such as the patent system may be superior.

With the procurement system, the government picks the firms that will be involved in the research project and specifies the terms of the project (such as expected research output and compensation terms) in a binding contract (Laffont and Tirole, 1993). Unlike the prize system, this method can eliminate unwanted duplication of research efforts, but again, for this system to be efficient, the government must be quite knowledgeable about the costs and benefits of research ventures.[7] More generally, innovations can be stimulated by the government's direct involvement in the production of new knowledge. Much of the research carried out at public institutions, and sponsored by public funds, is an example of such knowledge production. This structure, together with the complex social milieu characterizing academic institutions, has made possible the 'open science' environment that can take substantial credit for many scientific and technological breakthroughs (Stephan, 1996; David, 1998). The tension between the behavioural standards of open science and the privatization of new knowledge made possible by patents is readily apparent. The concern is that the increased reliance

[7] This issue may be less problematic when the government is the only intended customer of the innovation, as happens, for example, with research related to national defence.

on IPR mechanisms may be eroding the domain of public information and access to research tools, which could have serious long-term consequences for the vitality of the community of science (David, 2000).

Conclusions

The patent system has emerged as the central institution for asserting IPRs in many crucial fields of science and technology. From an economic point of view, patents offer a second-best solution to the market failure arising from the public-good nature of knowledge. As such, the patent system contributes to solving a problem but comes with shortcomings of its own, mostly because it creates market power positions that can adversely affect the economic performance of the system. In fact, for most of the 19th century, the patent system was under considerable criticism by the same economics profession that now provides the most valuable insights for its defence. This change is due to the increased appreciation for the critical role that innovations play in stimulating economic growth. The possibility of protecting discoveries through patents, for example, is credited for bringing about crucial technical improvements in the industrial revolution (Dutton, 1984).

As noted, the *ex post* inefficiency of the patent system is viewed as the necessary downside in providing enough inducement to undertake desirable R&D projects. The size of the inducement depends on the length and scope of the patent right. Ideally, such an inducement should be proportional to the cost of the R&D project, which means that the length, breadth and height of a patent should be tailored to each particular innovation. In addition to the cost of the project, such a tailored patent should also reflect the particular market conditions of the new product and/or process. Clearly, the patent system does not do that, and arguably it cannot do that. These limitations suggest that continued efforts are required to improve the workings of the patent system. A solid understanding of its complex (and sometimes subtle) economic implications, which we have tried to review in this chapter, should prove useful in this endeavour.

References

Arrow, K.J. (1962) Economic welfare and the allocation of resources for inventions. In: Nelson, R.R. (ed.) *The Rate and Direction of Inventive Activity: Economic and Social Factors*. Princeton University Press, Princeton, New Jersey.

Barton, J.H. (1998) Acquiring protection for improved germplasm and inbred lines. In: Erbisch, F.H. and Maredia, K.M. (eds) *Intellectual Property Rights in Agricultural Biotechnology*. CAB International, Wallingford, UK.

Barzel, Y. (1989) *Economic Analysis of Property Rights*, 2nd edn. Cambridge University Press, Cambridge.
Chang, H. (1995) Patent scope, antitrust policy and cumulative innovation. *RAND Journal of Economics* 26, 34–57.
Cornelli, F. and Schankerman, M. (1999) Patent renewals and R&D incentives. *RAND Journal of Economics* 30, 197–213.
David, P. (1998) Common agency and the emergence of 'Open Science' institutions. *American Economic Review – Papers and Proceedings* 88, 15–21.
David, P. (2000) A tragedy of the public knowledge commons? Stanford Institute for Economic Policy Research, Discussion Paper No. 00–02, October.
Dasgupta, P. and Maskin, E. (1987) The simple economics of research portfolios. *Economic Journal* 97, 581–595.
Dasgupta, P. and Stiglitz, J. (1980) Uncertainty, industrial structure and the speed of R&D. *Bell Journal of Economics* 1, 1–28.
DeBrock, L.M. (1985) Market structure, innovation and optimal patent life. *Journal of Law and Economics* XXVIII, 223–244.
Denicolò, V. (1996) Patent race and the optimal patent breadth and length. *Journal of Industrial Economics* XLIV, 249–265.
Dutton, H. (1984) *The Patent System and Inventive Activities During the Industrial Revolution 1750–1852*. Manchester University Press, Manchester.
Friedman, D.D., Landes, W.M. and Posner, R.A. (1991) Some economics of trade secret law. *Journal of Economic Perspectives* 5, 61–72.
Fudenberg, D., Gilbert, R., Stiglitz, J. and Tirole, J. (1983) Preemptive, leapfrogging and competition in patent race. *European Economic Review* 22, 3–31.
Gallini, N. (1992) Patent policy and costly imitation. *RAND Journal of Economics* 23, 52–63.
Gilbert, R. and Newbery, D. (1982) Preemptive patenting and the persistence of monopoly. *American Economic Review* 72, 514–526.
Gilbert, R. and Shapiro, C. (1990) Optimal patent length and breadth. *RAND Journal of Economics* 21, 106–112.
Green, J.R. and Scotchmer, S. (1995) On the division of profit in sequential innovation. *RAND Journal of Economics* 26, 20–33.
Heller, M.A and Eisenberg, R.S. (1998) Can patents deter innovations? The anticommons in biomedical research. *Science* 280, 698–701
Kamien, M.I. and Schwartz, N.L. (1974) Patent life and R&D rivalry. *American Economic Review* 64, 183–187.
Kitch, E.W. (1977) The nature and function of the patent system. *Journal of Law and Economics* 20, 265–290.
Klemperer, P. (1990) How broad should the scope of patent protection be? *RAND Journal of Economics* 21, 113–130.
Laffont, J.-J. and Tirole, J. (1993) *A Theory of Incentives in Procurement and Regulation*. MIT Press, Cambridge, Massachusetts.
La Manna, M. (1992) Optimal patent life vs. optimal patentability standard. *International Journal of Industrial Organization* 10, 81–90.
Lee, T. and Wilde, L. (1980) Market structure and innovation: a reformulation. *Quarterly Journal of Economics* 94, 429–436.
Loury, G. (1979) Market structure and innovation. *Quarterly Journal of Economics* 93, 395–405.

Machlup, F. and Penrose, E. (1950) The patent controversy in the nineteenth century. *Journal of Economic History* 10, 1–29.
Matutes, C., Regibeau, P. and Rockett, K. (1996) Optimal patent design and the diffusion of innovations. *RAND Journal of Economics* 27, 60–83.
Mazzoleni, R. and Nelson, R.R. (1998) The benefits and costs of strong patent protection: a contribution to the current debate. *Research Policy* 27, 273–284.
Merges, R.P. (1997) *Patent Law and Policy – Cases and Materials*, 2nd edn. Michie Law Publishers, Charlottesville, Virginia.
Merges, R.P. (1999) As many as six impossible patents before breakfast: property rights for business concepts and patent system reform. *Berkeley Technology Law Journal* 14, 577–615.
Merges, R.P. and Nelson, R.R. (1990) On the complex economics of patent scope. *Columbia Law Review* 90, 839–916.
Mowery, D.C., Nelson, R.R., Sampat, B.N. and Ziedonis, A.A. (2001) The growth of patenting and licensing by U.S. universities: an assessment of the effects of the Bayh–Dole Act of 1980. *Research Policy* 30, 99–119.
Nordhaus, W.D. (1969) *Inventions, Growth and Welfare: a Theoretical Treatment of Technological Change*. MIT Press, Cambridge, Massachusetts.
O'Donoghue, T., Scotchmer, S. and Thisse, J.-F. (1998) Patent breadth, patent life, and the pace of technological progress. *Journal of Economics and Management Strategy* 7, 1–32.
Reinganum, N. (1983) Uncertain innovation and the persistence of monopoly. *American Economic Review* 73, 741–748.
Reinganum, N. (1989) The timing of innovation. In: Schmalensee, R. and Willig, R.D. (eds) *Handbook of Industrial Organization*, Vol. I. North-Holland, Amsterdam, pp. 849–908.
Scherer, F.M. (1972) Nordhaus' theory of optimal patent life: a geometric reinterpretation. *American Economic Review* 62, 422–427.
Scotchmer, S. (1991) Standing on the shoulders of giants: cumulative research and the patent law. *Journal of Economic Perspectives* 5, 29–41.
Scotchmer, S. and Green, J. (1990) Novelty and disclosure in patent law. *RAND Journal of Economics* 21, 131–146.
Shapiro, C. (2001) Navigating the patent thicket: cross-licenses, patent pools, and standard-setting. In: Jaffe, A., Lerner, J. and Stern, S. (eds) *Innovation Policy and the Economy*, Vol. I. MIT Press, Cambridge, Massachusetts.
Stephan, P.E. (1996) The economics of science. *Journal of Economic Literature* XXXIV, 1199–1235.
Tandon, P. (1983) Rivalry and the excessive allocation of resources to research. *Bell Journal of Economics* 14, 152–165.
van Dijk, T. (1996) Patent height and competition in product improvements. *Journal of Industrial Economics* 44, 151–167.
Wright, B. (1983) The Economics of invention incentives: patents, prizes, and incentive contracts. *American Economic Review* 73, 691–707.

4

Access to Data and Intellectual Property

Margaret M. Dunbar

Cypress Bioscience, Inc., 4350 Executive Drive, San Diego, CA 92121, USA

Introduction

Access to data and generation of intellectual property (IP) related to animal genomics are evolving issues because computation science and bioinformatics are rapidly changing the face and nature of biological research. Computational science is a relatively new, but now essential component for successful animal genomics research. During the last 5 years or so, there has been a virtual explosion in the quantity of nucleic acid sequence data generated by and available to animal scientists. In addition, the use of 'DNA chips' for expression profiling has become more widespread as the technology becomes more accessible and the costs decrease. A series of experiments with a chip arrayed with several thousand different genes rapidly generates a staggering number of data points. The addition of sophisticated computer analysis to the researcher's toolbox allows for this otherwise overwhelming amount of data to be analysed and reduced to new discoveries. How these data and discoveries are best used and protected can be very important both to those who generate the data and those who wish to make use of this rapidly growing resource. Some of the issues surrounding the use of these data, the most common forms of IP protection along with some criticisms and alternatives are examined in this chapter.

In 1965, the founder of Intel Corporation, George Moore, observed that both computer chip performance and manufacturing capacity had doubled every 18–24 months. He also predicted that this doubling of performance would continue for the foreseeable future. 'Moore's Law', as this prediction has been dubbed, was remarkably accurate with the result that computing power has risen exponentially over a relatively brief period of time. More recently, there have been dramatic improvements in DNA sequencing technology that have contributed to large decreases in the cost of sequencing. This has resulted in an exponential increase in the number of DNA sequences being submitted to public databases. The rate at which new sequences were being submitted was estimated to be greater than 5000 per day (NIH, 1999) and is now certainly an even greater volume. Computers and computing speed have already transformed many sectors of business and society, and the explosion of sequence data now being generated is causing a comparable transformation in genomics as well. As little as a decade ago, genomics and computer science were both progressing more or less independently of one another. Yet, in the last few years, it has become clear that the continued growth in the use of genomics data is very dependent on the computer science disciplines, hence the emergence of the field of bioinformatics. Some 'genomics' technologies have also borrowed heavily from the computer industry. An example of which is some 'DNA chips'. The fabrication of some of these chips uses technology derived from the manufacture of microprocessors.

The continually expanding role of genomics in research has yielded more information, and holds greater potential for continued advances than originally projected. For example, the initial draft sequence of the human genome was finished in advance of its predicted schedule and the sequencing of other genomes is also proceeding ahead of most predictions made only a few years ago. This progress has resulted in data and information derived from that data being generated at a much faster rate than the general understanding of the different possible IP strategies.

Factors affecting Publication

Animal scientists continue to struggle to figure out the most useful and efficient ways to employ bioinformatics and computational science in their research. The increased use of computation in animal genomics has occurred so rapidly that many researchers have not had sufficient time to fully comprehend the advantages and possible disadvantages of sharing data. This is both from the perspective of publishing results derived from genomics research, providing access to their own data as well as using others' data. Because funding for research, whether in a commercial or academic setting, always seems to be in greater demand than supply, it is essential that the data generated by any institution provide optimal value. It is common in genomics research for a project or series of experiments to generate far more data than the individual or group can analyse in a meaningful amount of time. Whether to take this 'excess' data and publish, license, collaborate with others or simply hold on to it are options to be considered and exercised by the people generating genomic data as part of their research.

Some of these issues have already been recognized and studied. The National Institutes of Health (NIH) has several times published recommendations and comments acknowledging change in the role of genomics data in research. However, because of the speed of advancement in the field, even the recommendations of traditional leaders in providing guidelines, such as the NIH or leading peer-reviewed journals, are frequently changing and not in agreement with each other.

Initially any data generated from centres working on the Human Genome Project and receiving NIH funding were subject to a policy of release within 24 h (NHGRI, 1996). Later this policy was relaxed to requiring data release within 30 days (NHGRI, 1999). These policies have given way to guidelines that research involving genomic data should simply include some plan to disseminate the results, without a specific time constraint for the planned data release (NHGRI, 2000). Even though data is being generated at an unprecedented rate, the need to submit data for publication is considered to be less urgent now, as evidenced by the continued extension of time for data release. Even leading journals for reporting research, such as *Nature* and *Science*, differ in their requirements for access and release of data supporting research reported in their publications.

NIH has also studied and reported on the special problem of data and its use in its Biomedical Information Science and Technology Initiative, as well as in the NIH proposal for an Information Storage, Curation Analysis and Retrieval programme. The initiative recognizes the need to develop new types of databases that allow researchers to access, in a meaningful way, results from thousands of experiments simultaneously. This should encourage more rapid improvements in our understanding of genes, their functions and interactions (NIH, 1999).

In spite of the elimination of a within-24-h-data-release requirement, and a perception that it is less urgent that there be public access to data generated from genomics research, biologists have developed new expectations regarding access to data. Genomics researchers have become increasingly reliant on having ready access to shared data and analytical tools to enhance their research.

There is, however, a tension between keeping data proprietary because of the need to get value from one's own research, and the clear certainty that access to additional external data allows faster progress in disciplines such as genomics. This tension exists both at for-profit institutions, as well as universities and other non-profit organizations. There is a real risk that if data is held back in a very active area of research that in a relatively short amount of time, similar and overlapping data will be published by another group. Besides any IP considerations, in areas where funding is quite limited (e.g. agricultural animal species), failing to publish data in a timely fashion can lead to unnecessary duplication of effort and inefficient expenditure of limited resources.

This phenomenal volume of genomics data presents new challenges to making informed and rational decisions about the best use of the data from an IP perspective. The disciplines of bioinformatics and genomics are advancing so rapidly that there are unfilled positions for academic and industrial professionals in genomics, as well as open graduate student and post-graduate positions. Consequently the 'support' fields of technology transfer and IP counsel, which traditionally draw from this pool of professionals, has been slow to catch up and fill open positions with qualified personnel.

This lag in the existence of a pool of technology transfer personnel having a solid working knowledge in genomics is further exacerbating an existing knowledge gap regarding strategies and consequences of sharing huge volumes of computational data. It also increases the communication disconnect between researchers in the field and their liaisons in the relevant IP or technology transfer offices. In many instances, researchers in animal genomics are finding it difficult to not only get answers to their IP questions in this area, but sometimes the questions themselves are not even well appreciated by the offices and people the researchers need to rely on, as these collateral professions also struggle to learn bioinformatics and computational science.

None the less, even with these challenges, many researchers and their technology transfer counterparts have a good understanding of the traditional use of patents and licensing in animal science research. In some cases, this familiar framework will continue to serve, but in many cases these traditional tools will not be as useful as they have in the past.

Legislation affecting Genomics IP

A 1980 Supreme Court decision was a major factor influencing both progress and growth in the biotechnology field. In the case of *Chakrabarty vs. Diamond*, the Supreme Court ruled that manmade organisms were patentable resulting in Dr Chakrabarty being issued a patent for his 'oil-eating' bacteria. This decision set the precedent allowing patents, such as the patent issued to Cohen and Boyer (US patent no. 4,237,224), for a gene-splicing technique, as well as for patents on transgenic animals. The Cohen and Boyer patent was very successfully licensed for a small fee, and the process became a widely used research tool. About this same time, other policy changes were also taking place;

researcher is part of a non-profit institution, such as a university, they should take the time to inform themselves on the policy governing these issues. In many instances, a mandate to capture IP value from the research is not meant to block the path of the researcher pursuing research, but is supposed to be secondary or subservient to the pursuit and achievement of research progress. The better understanding a researcher has of their own institutional IP priorities, the better prepared they can be to work with their technology transfer colleagues.

If a researcher or their employer has as a top priority the capture of IP value from research, as is the case for much of the research pursued by for-profit institutions and industry, then their focus and attention in forming collaborations or exchanging data will be on the terms that best serve the capturing of the IP value from the data sharing. These goals are often embraced in future or automatic exclusive options and/or licensing rights in the resulting inventions in exchange for financial compensation. Being knowledgeable in advance of one's primary goals, and what IP rights or other concessions/value(s) one is willing to exchange to accomplish that goal, will help one to match up with the right partners and better articulate their goals in any sort of collaboration or other research endeavour.

This new high-volume world of genomics data seems to also call for new IP options. At larger institutions and companies that have a longer history with IP issues, it may be more difficult to gain acceptance for novel approaches. One less used option that is amenable to the vast, rapidly produced sets of genomics data is for one party to receive the option or rights to go forward with all the data in one area, such as kidney disease or ruminants and the other party to have the rights for all data relating to a different area, such as reproduction or primates. Similarly, if both parties are focused on the same problem or area, they may want to let one party get first pick to a small percentage of the results, with exclusive rights, and the other party to get second choice, but to the larger remaining percentage of data. These choices can be split 5:95, 10:90, etc., depending on the contributions of the collaborators.

In another alternative, a data-set owner can grant rights for a certain fixed amount of time or to the accomplishment of specific milestone advancement. If the conditions are not met, the owner then reclaims all of the rights in the data. In the case where the time limit and/or milestone is met, then the owner can consider requesting a return of right in any data that did not directly contribute to the specified outcome. As with every good IP agreement, it is very important that terms are well defined, so that the meaning of terms such as 'directly contribute' can be determined. Of overriding importance is the need to have IP terms that are frequently used in describing the use of genomics data, are at least understood by the parties, and ideally expressly defined in any agreement covering the data. If 'using' data and 'accessing' data are meant to mean the same thing in an agreement, then just one of the terms should be used exclusively to help keep the meaning clear. If accessing data means, for example, just 'looking' at data and 'using' the data means applying it in some way, then that should be clearly differentiated and stated as well. For example, if someone is just checking to see if a supposedly novel sequence exists in another's proprietary database, then it is important to the 'user' or 'accessor' to know whether to expect to acquire obligations when no match is found and if the data-set owner has the same expectations. The more valuable any result is, the more likely that the parties will potentially disagree on any terms that are not clear. A clear understanding in the data IP terms in an agreement can prevent the breakdown of a relationship between otherwise well-suited research collaborators.

In the case of multiple groups exchanging and sharing data, as the number of participants increases the likelihood that the rights and data will be all shared equally or be dedicated to the public increases. Under most situations a large group of separate collaborators are not going to agree that one or two of the group should have disproportionate rights, unless there is some special justification for the difference.

Conclusions

It appears that there are two different kinds of chimeras that can arise out of animal genomics today. There are chimeric animals that hold the promise of advancement of knowledge and hope for medical progress, both for humans and animals, and there is the data chimera produced from the fusion of computational science and the study of genetics, which also hopefully holds forth the promise of great progress in animal genomics.

Despite the challenges that can arise, researchers should continue to let others use their data as well as arranging to bring others' data into their own research. These activities can be accomplished under a variety of terms, ideally in alignment with each researcher's goals or as prescribed by their employers. Most importantly, they should be informed of their options in this regard as well as the potential problem areas, so that they can avoid the foreseeable problems and, to the extent possible, have a plan when confronted by the unforeseen problems that may arise from sharing data.

References

Benbrook, C. (2000) Who controls and who will benefit from plant genomics? AAAS Annual Meeting, http://www.biotech-info.net/AAASgen.html

Clark, J., Piccolo, J., Stanton, B. and Tyson, K. (2000) Patent pools: a solution to the problem of access in biotechnology patents? United States Patent and Trademark Office, pp. 2–11.

Clinton, W. and Blair, T. (2000) Joint Statement by President Clinton and Prime Minister Tony Blair of the UK. http://clinton4.nara.gov/WH/EOP/OSTP/NSTC/html/bioinformaticsreport.html

National Human Genome Research Institute (NHGRI) (1996) NHGRI Policy Regarding Intellectual Property of Human Genomic Sequence.

National Human Genome Research Institute (NHGRI) (1999) Policy on Availability of Genomic DNA Sequence Funded by NHGRI.

National Human Genome Research Institute (NHGRI) (2000) Update of NHGRI Policy For Release and Database Deposition of Sequence Data.

NIH Working Group on Biomedical Computing (1999) The Biomedical Information Science and Technology Initiative.

5

Global Intellectual Property: International Developments in Animal Patents

Andrew Blattman, John McCann, Colin Bodkin and Jovanka Naumoska

Spruson & Ferguson, Patent and Trade Mark Attorneys, Level 35/31 Market Street, Sydney, NSW 2000, Australia

Introduction

The principles behind the patenting of new inventions remain essentially the same today as they were centuries ago. For instance, Section 6 of the English Statute of Monopolies of 1623 is expressly preserved in the Australian Patents Act 1990.[1] A challenge for the patent systems of the world is to apply the established principles, in particular those relating to the definition of patentable subject matter, to fit today's technological advances.

While inventions made in the mechanical field have been absorbed into the patent system with little difficulty, there has been a greater uneasiness associated with the granting of patent protection to inventions relating to biological entities and processes and in particular, to higher life forms. In this regard, a survey conducted by the World Intellectual Property Organisation (WIPO) in 1988 found at least 45 countries throughout Europe, Africa and Asia which prohibited patents for animal varieties (WIPO, 1988).

This chapter is in two parts. The first part examines the international cooperation that exists with respect to obtaining patent protection for inventions, and the second part specifically focuses on the national positions with respect to patent protection for animal inventions in the following jurisdictions: USA, Canada, Europe and Australia, with a short discussion on the positions in New Zealand, South Africa, Japan, China, Brazil and Argentina.

[1] See definition of 'invention' in Schedule 1 of the Australian Patents Act 1990.

The question of the patentability of animal inventions is not merely answered by considering the patentability of animals *per se*. Straus (1989) has suggested that in considering animal inventions there are at least five distinct categories of invention:

1. Animals and animal varieties *per se*;
2. Animals and animal varieties as products of a process;
3. Processes for the production of animals;
4. Animal materials *per se*; and
5. Processes for the production of animal materials.

Generally, countries of the world have found little difficulty in recognizing innovations in categories 4 and 5 as patentable. The discussion of patentability of animal-related inventions in this chapter will therefore focus on inventions in categories 1–3. We turn first, however, to a consideration of international arrangements relating to patent protection.

International Intellectual Property (IP) Conventions

General

Nearly all major animal breeding countries are signatories to a number of international IP treaties. The national IP legislation of many of these countries contain provisions implementing these treaties and incorporating their undertakings into domestic law. As far as patents are concerned, the major international treaties which affect animal patents are as follows:

1. Paris Convention for the Protection of Industrial Property of 1883 (Paris Convention);
2. The Patent Cooperation Treaty 1970 (PCT); and
3. Agreement on the Trade Related Aspects of Intellectual Property Rights (TRIPs).

Paris Convention

The Paris Convention was formed in 1883, and came into effect on 7 July 1884. It is probably the single most important multi-lateral treaty relating to IP, and in particular relating to patents, trade marks, service marks, utility models, etc. Since its inception, the Paris Convention has gone through periodic revisions (1900, 1911, 1925, 1934, 1958 and 1967), the latest being the Stockholm text of 1967. As of 15 January 2002, there were 162 signatory countries to the Paris Convention.

An important provision of the Convention is set out in Article 2(1), which requires member countries to accord the same right to nationals of other Union countries as it grants to its own nationals. This treatment is witnessed for example, wherein a person filing an application for a patent in any Convention country may file an application for a patent in any other Convention country within 12 months of the date of first filing without loss of priority.[2] This right is known as the right of priority.

PCT

As outlined above, the main function of the Paris Convention is to provide a means of access into different countries' patent systems. However, once lodged, a patent application must be prosecuted through each national patent system, a procedure involving considerable duplication and expense. Therefore, the PCT was entered into in order to provide patent applicants with a relatively simple way of commencing patent applications in a number of countries simultaneously.

The PCT was first signed in Washington, DC, in 1970, and came into effect on 24 January 1978. As of January 2002, there were 115 signatories to the PCT, and the list continues to grow. Most major trading countries are signatories, with the exceptions including many South American countries (except Brazil, Costa Rica and Guinea), Taiwan, Malaysia, Pakistan and Thailand. South Africa, Indonesia, Philippines and India are all recent additions.

Essentially, a single (PCT) application is filed at a national patent office, so that the patent applicant only has to meet a single set of formality requirements. This single application contains designations of the relevant states and any priority claims. The PCT procedure is split into two parts. The first part relates to International Searching and Publication, whilst the second part relates to international preliminary examination.

Once a PCT application has been filed, a national patent office, acting as an international searching authority, conducts a prior art search in order to establish the international search report, which is sent to the applicant. In response, an applicant may amend the claims and optionally file a statement (500 words maximum related to the amendments), but is not able to amend the specification.

The International Bureau of WIPO publishes the PCT application, and this generally occurs 18 months from earliest priority date. The 'national phase' is then ready to begin. The 'national phase' refers to commencement of processing the patent application separately in each national patent office where the patent applicant ultimately desires to obtain a patent.

---

[2] Paris Convention for the Protection of Industrial Property, Article 4.

If the applicant wishes to take advantage of international preliminary examination, then it is necessary to lodge a 'demand' for international preliminary examination with WIPO before the expiry of the 19th month reckoned from the priority date. In international preliminary examination, a preliminary and non-binding opinion (written opinion) is formulated on issues of novelty, inventive step and whether the invention is thought to be industrially applicable. The applicant may respond to a written opinion by amending both the claims and the specification, and/or by filing argument.

In the past, assuming a demand was lodged in good time before the expiry of the 19th month, the deadline for entry into the national phase was extended to 30 months (31 months in some countries). However, effective 1 April 2002, the PCT has been amended to remove the requirement that a demand be filed prior to the 19th month deadline to postpone national phase entry.

This change will take effect in those member states which are able to amend their national legislation by the 1 April 2002 deadline. Under transitional arrangements, where national legislation is incompatible with the changed time limits, the time limit for national phase entry remains at 20 or 21 months (in the absence of a demand being filed).

TRIPs

The TRIPs agreement came into effect on 1 January 1995, and is part of the Marrakesh Agreement establishing the World Trade Organisation (WTO). TRIPs establishes multilateral international standards for the protection of IP rights. In essence, the TRIPs agreement requires countries which are members of the WTO to comply with the substantive obligations of the most recent versions of the Paris and the Berne (concerning copyright) Conventions through incorporating the substantive provisions of these conventions via cross-reference (except for the provisions on moral rights in the Berne Convention). The aims of the TRIPs agreement were to reduce distortions and impediments to international trade, to provide effective and adequate protection of IP rights (IPRs), and to ensure that measures and procedures to enforce IPRs do not themselves become barriers to legitimate trade.

The areas of IP covered by TRIPs are copyright and related rights, trademarks including service marks, geographical indications including appellations of origin, industrial designs, patents including the protection of new varieties of plants, the layout designs of integrated circuits, and undisclosed information including trade secrets.

In terms of patent protection relating to animals, the TRIPs agreement requires member countries to make patents available for any inventions, whether products or processes, in all fields of technology without discrimination, subject to the normal tests of novelty, inventiveness (non-obviousness) and industrial applicability.

As far as the patentability of an animal invention is concerned a crucial TRIPs provision is provided in Article 27.3 reproduced below.

> 27.3 Members may also exclude from patentability:
> (a) diagnostic, therapeutic and surgical methods for the treatment of humans or animals;
> (b) *plants and animals other than micro-organisms, and essentially biological processes for the production of plants or animals* other than non-biological and microbiological processes. However, Members shall provide for the protection of plant varieties either by patents or by an effective *sui generis* system or by any combination thereof. The provisions of this subparagraph shall be reviewed four years after the date of entry into force of the WTO Agreement. (Emphasis added)

Whilst the TRIPs agreement provides a minimum standard that has to be incorporated and enforced by each member country, each member country is free to determine the appropriate method of implementing the provisions of the agreement within its own legal system and practice. Hence, even though member countries may implement the same standard of protection, the differing applications of this standard within their legal systems has created differing practices and procedures.

TRIPs is directly relevant to the question of animal patents since Article 27.3 provides member countries with the option to exclude from patentability animals other than microorganisms, and essentially biological processes for the production of animals, and some countries have indeed incorporated these exclusions into their national patent legislation.

Animal Patenting Positions Around the World

The following sets out the position with respect to animal patents in a number of countries and regions, with specific reference to the USA, Canada, Europe and Australia, together with a brief commentary on the respective positions in New Zealand, South Africa, Japan, China, Brazil and Argentina.

USA

In the USA, criteria governing the patentability of any innovation are set out in Section 101 of Title 35 of the US Code (U.S.C.). Title 35 U.S.C. §101 provides as follows:

> Whoever invents or discovers any new and useful process, machine, manufacture, or composition of matter, or any new and useful improvement thereof, may obtain a patent therefore, subject to the conditions and requirements of this title.

Title 35 U.S.C. §101 thus confers patentability, at least potentially, on methods for generating or breeding animals and, if animals come within the meaning of 'composition of matter', on animals themselves, subject of course to the animal satisfying the 'newness' requirement of 35 U.S.C. §101 and other requirements of the statute.

The Congressional Committee Reports that accompanied the current legislation when it was enacted stated that the Congress intended 'anything under the sun that is made by man' to be capable of patent protection (Senate Report, 1952). In this context, in the case of *Diamond vs. Chakrabarty* (1980a), the US Supreme Court ruled that a genetically engineered microorganism was patentable.

Diamond vs. Chakrabarty concerned a patent for a bacterium that had been genetically engineered to degrade crude oil. The US Patent Office Board of Appeals had declined to grant Chakrabarty's claims to the genetically modified microorganism on the ground that living things were not patentable under 35 U.S.C. §101. The rejection was reversed by the US Court of Customs and Patent Appeals, whereupon the Commissioner of Patents and Trademarks brought the case before the Supreme Court of the USA. By majority, the Supreme Court held that the bacterium was 'a nonnaturally occurring manufacture or composition of matter – a product of human ingenuity having distinctive name, character and use' (*Diamond vs. Chakrabarty*, 1980b), and that the patentee had 'produced a new bacterium with markedly different characteristics from any found in nature and one having the potential for significant utility' (*Diamond vs. Chakrabarty*, 1980c). The Supreme Court observed that 'Congress is free to amend §101 so as to exclude from patent protection organisms produced by genetic engineering ... (o)r it may choose to craft a statute specifically designed for such living things. But, until Congress takes such action, this Court must construe the language of §101 as it is. The language of that section fairly embraces respondent's invention.'

The applicability of the Chakrabarty decision to the patentability of higher life forms was tested in the decision of *Ex Parte Allen* (2 U.S.P.Q.2d (BNA) 1425), which concerned a patent for oysters that had been made more edible by pressurizing them. The Patent Examiner had rejected the patent application on the ground, *inter alia*, that it was directed to a naturally occurring organism. On appeal, the Board of Patent Appeals, although it ultimately rejected the patent on the ground of obviousness, considered that it was not a bar to patentability that the claims were directed to a multicellular organism. The issue, the Board ruled, was not whether the subject matter was a product of nature, but whether it was made by man (2 U.S.P.Q.2d (BNA) 1427).

Soon after the decision in *Ex Parte Allen* the US Patent and Trademark Office (US PTO) issued an announcement that:

> The Patent and Trademark Office now considers non-naturally occurring non-human multicellular living organisms, including animals, to be patentable subject matter within the scope of 35 U.S.C. §101.

The USPTO also stated that, in order for an animal to be patentable, it must be 'given a new form, quality, properties or combination not present in the original article existing in nature in accordance with existing law' (1077 OG 24).

Since the USPTO's announcement, a number of bills have been introduced into the Congress for the purpose of regulating the patenting of animals. For example, HR 3119, introduced on 5 August 1987 by Congressman Rose and others, proposed a 2-year moratorium on granting patents on animals, to allow Congress to discuss the relevant issues involved in animal patenting. A further bill (S 2111, introduced on 29 February 1988 by Senator Hatfield) contained measures to ban animal patents completely. However, neither these bills nor any other measure concerned with animal patents has been passed by both houses of Congress.

Within a year of the USPTO's announcement, patent no. 4,736,866 was granted to Leder and Stewart of Harvard University. This was the first patent of its kind to be granted in that it claimed a non-human mammal. It has come to be known as the 'Harvard mouse' patent. Claim 1 is:

> 1. A transgenic non-human mammal all of whose germ cells and somatic cells contain a recombinant activated oncogene sequence introduced into said mammal, or an ancestor of said mammal, at an embryonic stage.

The claim thus covers animals other than mice (but not including humans), and also covers the descendants of the animal into which the oncogene sequence was first introduced.

Although a number of animal rights groups sought to challenge the right of the USPTO to grant patents for animals, the District Court and the Court of Appeals for the Federal Circuit held that these groups had no standing to do so. At the time of writing, it remains the case that no US court has ruled on the patentability of transgenic animals.

Since the granting of the Harvard mouse patent, the USPTO has granted numerous patents directed to transgenic animals, including transgenic sheep, pigs, goats and cattle. The patents are directed either to transgenic animals as models for some human disease, or to animals that have been modified genetically to produce useful proteins.

A more recent development has been the granting, by the USPTO, of patents directed to methods of cloning animals. The first such patent to have been granted appears to have been US patent no. 5,945,577 granted on 31 August 1999 to Stice *et al.* of the University of Massachusetts. Patent nos 6,235,969 and 6,235,970, also to Stice *et al.*, are related. The patents to Stice *et al.* derive from a patent application first filed on 10 January 1997. In a similar technical field are two patents to Campbell and Wilmut, claiming priority from a UK patent application filed on 31 August 1995 and granted on 14 November 2000 (US patent no. 6,147,276) and 26 June 2001 (US patent no. 6,252,133), respectively. The patentees are part of the team of researchers that announced the successful cloning of 'Dolly' the sheep in February 1997. Claim 1 of patent no. 6,147,276 reads:

> 1. A method of reconstituting a non-human mammalian embryo, comprising
>   a. transferring the nucleus of a quiescent diploid donor cell into a suitable enucleated recipient cell of the same species, thereby obtaining a reconstituted cell;
>   b. activating the recipient cell before, during or after nuclear transfer;
>   c. incubating the reconstituted cell such that the embryo develops.

The patent also includes claims to methods of preparing non-human mammals comprising reconstituting an embryo according to the method of claim 1, transferring it to a female of the same species and developing the embryo into the non-human mammal.

A further patent claiming a method of cloning animals was granted to Wakayama and Yanagimachi of the University of Hawaii on 18 December 2001 (US patent no. 6,331,659).

None of the patents to Campbell and Wilmut, to Stice *et al.* or to Wakayama and Yanagimachi includes a claim to a non-human mammal produced by the claimed method. The reason is not clear, but the present authors speculate that claims to such animals may have been rejected as failing the 'newness' requirement. However, since animals produced by the claimed methods are not in fact identical to either the oocyte donor or to the animal from which the donor cell was obtained, it seems to be arguable that they are demonstrably distinct from the 'prior art' animals.

It has been noted above that the USPTO considers that humans are not patentable subject matter. Although there is no explicit authority for this practice in Title 35 U.S.C., the practice is consistent with statutory exclusions to patenting humans that exist in numerous other jurisdictions. The USPTO has based its practice on the anti-slavery provisions found in the 13th Amendment to the US Constitution. In any event, it may be difficult to argue in most cases that a claim that encompasses, say, a transgenic human being satisfies the 'usefulness' requirement of 35 U.S.C. §101.

Canada

Case law and patent office practice in Canada are less well developed than in the USA, no doubt because of the longer period between filing an application and its examination that is typical of patent applications in Canada, compared with the USA. In Canada, interest in this area has centred on the progress of the 'Harvard mouse' application, but the final decision by the Canadian courts on the patentability of the oncomouse is still awaited.

In Canada, in order to be patentable, an innovation must qualify as an 'invention', the definition of which is very similar to the US requirements for patentability. Thus, section 2 of the Canadian Patent Act 1985 defines 'invention' as 'any new and useful art, process, machine, manufacture or composition of matter, or any new and useful improvement in any art, process, machine or composition of matter'.

In *Re Application of Abitibi Co.* (1982), the Patent Appeal Board considered a patent application claiming a yeast culture, which, although naturally occurring, had to be acclimatized and treated to enable it to work on spent sulphite waste. The Board held that the claim was patentable, and recognized that the decision extended to 'all new life forms which are produced en masse as chemical compounds are prepared, and are formed in such large numbers that any measurable quantity will possess uniform properties and characteristics'. The Board considered it to be more debatable whether its decision extended to higher life forms such as plants and animals.

Subsequently, the question of the patentability of plants arose in *Pioneer Hi-Bred vs. Canada (Commissioner of Patents) (1987)*, in which the Federal Court of Appeal rejected the patent application *inter alia* on the ground that the words 'manufacture' and 'composition of matter' did not include a variety of soybean within their scope. The Supreme Court upheld the decision of the Court of Appeal on the ground that the disclosure of the application was insufficient, and did not consider whether plants were patentable *per se*. The issue of the patentability of higher life forms in Canada thus remained open after *Pioneer Hi-Bred*.

In this context, the 'Harvard mouse' application came to be examined by the Canadian Patent Office. The patent examiner allowed the claims to the process for obtaining the transgenic animal, but rejected the claims directed to the transgenic animal itself, taking the view that the patentability of 'higher life forms' was prohibited in Canada. On appeal to the Commissioner of Patents in 1995, the examiner's rejection was upheld. The Commissioner stated that if the oncomouse invention was divided into two phases, the first being the genetic engineering of the plasmid with the oncogene and its introduction into the mouse genome, and the second being the actual development of the genetically engineered mouse in the uterus of the host mouse, then it can be said that the former was controlled by human intervention, and therefore was patentable, while the latter was controlled by the laws of nature, and was therefore not patentable. Further, the Commissioner stated that the element of reproducibility was not satisfied since the inventors did not have full control over all the characteristics of the mouse, only of the oncogene.

In 1998, Harvard appealed the decision of the Commissioner to the Trial Division of the Federal Court of Canada. Mr Justice Nadon, refusing the appeal, held that 'higher' life forms were not patentable subject matter (*President and Fellows of Harvard College vs. Canada*, 1988):

> A complex life form does not fit within the current parameters of the Patent Act without stretching the meaning of the words to the breaking point, which I am not prepared to do. However, if Parliament so wishes, it clearly can alter the legislation so that mammals can be patented.

The appellants appealed to the Federal Court of Appeal, which, by majority, overturned the decisions of the Trial Division and the Commissioner. The Court of Appeal referred to the decision of the Supreme Court of the USA in *Chakrabarty* and stated that the '[t]he language of patent law is broad and general and is to be given wide scope because inventions are, necessarily, unanticipated and unforeseeable' (*President and Fellows of Harvard College vs. Canada*, 2000). The Court, in reaching its decision, placed much weight on the reasoning in *Chakrabarty* and considered that '[the respondent's] claim is not to a hitherto unknown natural phenomenon, but to a non-naturally occurring manufacture or composition of matter – a product of human ingenuity having distinctive name, character and use.'

The Court also relied on the fact that the patent examiner had determined that the 'process' claims were patentable. It observed:

> If the process for producing the product is patentable, it is because it must be considered to involve ingenuity and not merely the discovery of the operation of a law of nature or the existence of a naturally occurring phenomenon. It must logically follow that the product of that process must also be considered to involve that same ingenuity and be patentable.

Since the decision of the Federal Court, the Canadian government (with the Commissioner of Patents as applicant) has appealed to Canada's highest court – the Supreme Court of Canada. The appeal was heard on 21 May 2002.[3] The decision of the Supreme Court has not yet been handed down. Hence, although it is clear that processes for producing animals are capable of patent protection in Canada, a decision of the Supreme Court must be awaited before the question of whether or not higher life forms such as animals are patentable is finally settled. It has been reported that a backlog of patent applications for transgenic animals in Canada awaits the final word from the Court on this issue.

At the time of writing, changes to the Patent Act 1985 are under consideration by the Canadian government. The Canadian Biotechnology Advisory Committee, an independent advisory group that was established by the Canadian government to advise it on biotechnology policy issues, has issued an interim report on patenting of

[3] Case No. 28155.

higher life forms and related issues. The report includes recommendations that higher life forms that meet the criteria of novelty, non-obviousness and utility should be recognized as patentable in Canada, subject to certain limitations as to infringement and liability for damages (Canadian Biotechnology Advisory Committee, 2002). The report was presented to the Government of Canada in June 2002.

Europe

In Europe, essentially two different routes can be pursued for obtaining patent protection in one or more European countries: specifically, the national application filing route and the European patent application filing route. In the national application filing route, a patent application is filed in each of the patent offices of the European countries where patent protection is sought. Each of those patent offices separately determines, via an examination process, whether to refuse the application or whether to accept the application and thereafter to grant a patent on it. In the European patent application filing route, which is made possible by the European Patent Convention[4] (EPC), one European patent application which designates the member European countries[5] where patent protection is sought is filed at the European Patent Office (EPO) and a substantial examination process within the EPO follows. The EPO determines whether to refuse the application or whether to accept the application and thereafter to grant a single European patent on it. Once granted, the European patent is validated in some or all of the designated European countries. Post-patent grant opposition is available to third parties who may file a notice of opposition at the EPO within 9 months of the grant of a European patent.

The main advantage of proceeding via the European patent route is that there is a single examination process at the EPO as compared to multiple examination processes when national filings are pursued in two or more European countries. Most European countries are signatories to the EPC and consequently their patent laws have been harmonized with it. Whilst the EPC and some recent relevant developments in the EPO are discussed below, it is beyond the scope of this work to consider the variations in approaches adopted by the national Patent Offices of individual European countries.

[4] Convention on the Grant of European Patents.

[5] There are currently 24 signatories to the EPC, namely Austria, Belgium, Bulgaria, The Czech Republic, Estonia, Switzerland/Liechtenstein, Cyprus, Germany, Denmark, Slovakia, Spain, Finland, France, UK, Greece, Ireland, Italy, Luxembourg, Monaco, Netherlands, Portugal, Sweden and Turkey. There are also six states for which a European patent can be extended, namely Albania, Lithuania, Latvia, former Yugoslav state of Macedonia, Romania and Slovenia.

Article 52[6] of the EPC sets out the general requirements for patentability of inventions as well as listing non-patentable matter. In addition, Article 53[7] of the EPC states that European patents shall not be granted for a number of invention types.

On 1 September 1999, new rules[8] of the EPC came into effect at the EPO following the taking up of European Directive on the Legal Protection of Biotechnological Inventions (Directive 98/44/EC) in 1998, which directed that the signatories to the EU harmonize their patent laws in respect of the patenting of biotechnological subject matter.

Rules 23(b)(2), 23(b)(3), 23(b)(5) and 23(b)(6) of the EPC which came into being as a result of the European Directive (Directive 98/44/EC) respectively define biotechnological inventions, biological material, essentially biological and microbiological process as:

> (2) 'Biotechnological inventions' are inventions which concern a product consisting of or containing biological material or a process by means of which biological material is produced, processed or used.
> (3) 'Biological material' means any material containing genetic information and capable of reproducing itself or being reproduced in a biological system.
> (5) A process for the production of plants or animals is essentially biological if it consists entirely of natural phenomena such as crossing or selection.
> (6) 'Microbiological process' means any process involving or performed upon or resulting in microbiological material.

---

[6] (1) European patents shall be granted for any inventions which are susceptible of industrial application, which are new and which involve an inventive step.
(2) The following in particular shall not be regarded as inventions within the meaning of paragraph 1:
(a) discoveries, scientific theories and mathematical methods;
(b) aesthetic creations;
(c) schemes, rules and methods for performing mental acts, playing games or doing business, and programmes for computers;
(d) presentations of information.
(3) The provisions of paragraph 2 shall exclude patentability of the subject matter or activities referred to in that provision only to the extent to which a European patent application or European patent relates to such subject matter or activities as such.
(4) Methods for treatment of the human or animal body by surgery or therapy and diagnostic methods practised on the human or animal body shall not be regarded as inventions which are susceptible of industrial application within the meaning of paragraph 1. This provision shall not apply to products, in particular substances or compositions, for use in any of these methods.

[7] (a) inventions the publication or exploitation of which would be contrary to 'ordre public' or morality, provided that the exploitation shall not be deemed to be so contrary merely because it is prohibited by law or regulation in some of the Contracting States;
(b) plant or animal varieties or essentially biological processes for the production of plants or animals; this provision does not apply to microbiological processes or the products thereof.

[8] Rules 23b, 23c, 23d and 23e of the Implementing Regulations to the Convention on the Grant of European Patents.

Rules 23c(a)–(c) of the EPC state that the following biotechnological inventions are also patentable provided they concern:

> (a) biological material which is isolated from its natural environment or produced by means of a technical process even if it previously occurred in nature;
> (b) plants or animals if the technical feasibility of the invention is not confined to a particular plant or animal variety;
> (c) a microbiological or other technical process, or a product obtained by means of such a process other than a plant or animal variety.

Rule 23d(d) provides the following exception to patentability of animals:

> (d) processes for modifying the genetic identity of animals which are likely to cause them suffering without any substantial medical benefit to man or animal, and also animals resulting from such processes.

Consequences that follow from the above Articles and Rules as far as the patentability of animals under the EPC is concerned and that are noteworthy include:

1. Biotechnological inventions concerning biological material (even if they previously occurred in nature), currently including biotechnological inventions directed to genetically transformed animals, are patentable under the EPC, provided they meet the criteria of patentability specified by the relevant articles and rules of the EPC, even if that biological material has been isolated from its natural environment or produced by a technical process.

Thus, the Harvard Oncomouse patent application (EP Application No. 85304490.7), which concerned a genetically engineered transgenic mouse that had a susceptibility to developing cancer, was refused initially by the Examining division of the EPO on the basis that it claimed an animal variety, and was thus contrary to Article 53(b). The refusal was appealed (T19/90), and the Board of Appeal found that while Article 53(b) precluded specific categories of animals from patentability, it did not preclude patentability of broad classes of animals, and returned the application back to the Examining division with specific instructions to consider Article 53(b), weighing up the suffering of the animal against the potential benefits to humanity. Consequently, a European patent was granted on the application (European Patent No. 0169672) in respect of oncomammals. However, the patent was then opposed by 17 opponents, including Greenpeace. The Opposition Division of the EPO recently upheld the patent only in respect of oncorodents and not in respect of other oncomammals, but it is believed that this decision will be appealed by the British Union for the Abolition of Vivisection.

Consistent with the EPO's approach in the above-discussed Harvard Oncomouse case, a recent decision (LELAND STANFORD/Modified animal, 2002a) of the Opposition Division of the EPO allowed a claim directed to a chimeric non-human immunocompromised mammalian host[9] implanted with human haematopoietic tissue derived from children under 3 years or aborted fetuses. The claim was allowed to non-human animals rather than mice because the opponents did not satisfy the Opposition Division that animals other than mice could not be successfully used.

2. General classes of known animals (and plants), as opposed to specific animal (and plant) varieties, are now not precluded from patentability under the EPC, provided the technical feasibility of the invention is not limited to a particular animal (or plant) variety and the claims of the patent are directed to a general class of animals (or plants) which have been genetically modified by an essentially non-biological process even when the scope of such claims include animal (or plant) varieties.[10] In this connection, the EPO recently granted a European patent having claims directed to a transgenic plant,[11] although this patent was opposed. The subsequent decision by the Enlarged Board of Appeal of the EPO decided that a claim which embraces plant varieties but does not specifically claim individual plant varieties is not excluded from patentability under Article 53(b).[12]

3. The definition of 'essentially biological' in Rule 23(b)(5) is restricted to 'entirely natural phenomena such as crossing or selection'. It follows that the preclusion from patentability of 'essentially biological processes for the production of plants or animals' specified in Article 53(b) will be avoided if the process involves genetic modification of the animal by a non-natural biotechnological procedure (e.g. by DNA microinjection, retrovirus infection or embryonic stem cell gene transfer techniques). Furthermore, it is worth bearing in mind

---

[9] 1. A chimeric non-human mammalian host comprising:

an immunodeficient mammalian host comprising at least a portion of an immune system, lacking functional syngeneic lymphocytes but in which the thymus is present; and

wherein at least one organ comprises a xenogeneic functional thymus and xenogeneic cells and, optionally, xenogeneic tissues and/or organs, said cells comprising at least one of hematopoietic stem cells, immature differentiated hematopoietic cells or mature differentiated hematopoietic cells, as a result of introduction of said xenogeneic cells, tissue and/or organ into said mammalian host at an age of at least neonate.

[10] Directive 98/44/EC, *supr.* The following decisions are also relevant: HARVARD/Transgenic Mouse, *supra*, *Ciba-Geigy/Propagating Material* [1979–85] EPOR 758, *Hybrid Plants/LUBRIZOL*, O.J. EPO 1990, 71, *Plant Genetic Systems/Glutamine Synthetase Inhibitors* O.J. EPO 1995, 545, Decision T1054/96, *Plant varieties/NOVARTIS*, and Decision of the Enlarged Board of Appeal of the EPO issued 20 December, 1999 under case G1/98, *Plant varieties/NOVARTIS.*

[11] *Plant varieties/NOVARTIS, supra.*

[12] Decision of the Enlarged Board of Appeal of the EPO, *Plant varieties/NOVARTIS, supra.* This decision while consistent with Directive 98/44/EC, *supr.* did not rely on it.

that 'this provision does not apply to microbiological processes or the products thereof'. Thus, claims 1 and 21 of granted European patent no. 0 849 990, for example, respectively define a method reconstituting a non-human mammalian embryo and a method of preparing a non-human mammal.[13]

4. When considering whether an invention is patentable, the EPO is required to take into account ethical considerations as to whether it is precluded from patentability on the basis that it is contrary to *ordre public* or morality.[14] In the HARVARD/Oncomouse case[15] not only was due consideration given by the EPO to whether the process for modifying the genetic identity of the mice was likely to cause them suffering without any substantial medical benefit to man or animal, but the public interest in remedying cancer as well as the prevention of any unnecessary spread of undesired genes which increased susceptibility to cancer were also important considerations in the process of deciding that the transgenic mice were patentable.

In the LELAND STANFORD/Modified Animal case where it was agreed by the parties that there '... are enormous medical benefits associated with the invention' (LELAND STANFORD/Modified Animal case, 2002b) since the chimeric animals provide an animal model for infection with HIV-I the Opposition Division of the EPO found that:

(a) '... [the] argument that the patenting of animals is regarded as intrinsically unethical in Western society cannot ... be followed', in view of Rule 23d(d) of the EPC (LELAND STANFORD/Modified animal, 2002c); and
(b) '... as long as a claimed invention has a legitimate use, it cannot be the role of the EPO to act as a moral censor and invoke the provisions of Article 53(a) EPC to refuse on ethical grounds to grant a patent on legal research and directed to an invention indisputably associated with medical benefits' (LELAND STANFORD/Modified animal, 2002d).

---

[13] 1. A method of reconstituting a non-human mammalian embryo, comprising
(a) transferring the nucleus of a quiescent diploid donor cell into a suitable enucleated recipient cell of the same species, thereby obtaining a reconstituted cell;
(b) activating the recipient cell before, during or after nuclear transfer;
(c) incubating the reconstituted cell such that an embryo develops.
21. A method of preparing a non-human mammal, comprising:
(a) reconstituting a non-human mammalian embryo according to the method of claim 1; and
(b) transferring the embryo to a female of the same species;
(c) developing the embryo into the non-human mammal.

[14] HARVARD/Transgenic Mouse, *supra*, *Ciba-Geigy/Propagating Material, supra*, *Hybrid Plants/LUBRIZOL, supra*, *Plant Genetic Systems/Glutamine Synthetase Inhibitors, supra*, and Decision T0356/93, PGS/GREENPEACE.

[15] HARVARD/Transgenic Mouse, *supra.*

Australia

In Australia, patent protection is available under the Australian Patents Act 1990, wherein Section 18 thereof essentially indicates what is patentable.[16] To be patentable in Australia, an invention must in essence: be a 'manner of manufacture', be novel, involve an inventive step, be useful and not have been secretly used before the priority date. Finally, an invention must not transgress the limitation of section 18(2), in terms of being an invention directed to human beings *per se*, nor the biological processes for their generation.

Over the years, the Courts have sought to define what is meant by 'manner of manufacture'. In one of its leading decisions with respect to patents, the High Court of Australia, in *National Research and Development Corporation vs. Commissioner of Patents* (1959a), (hereinafter '*NRDC*'), adopted a wide approach to determining the nature of 'manner of manufacture'. In essence, the High Court found that to be a manner of manufacture, it was sufficient that a process 'offer some advantage which is material, in the sense that the process belongs to a useful art as distinct from a fine art – that its value to the country is in the field of economic endeavour' (*NRDC,* 1959b).

There is nothing in *NRDC* which is inconsistent with the patenting of animals, although the question is yet to receive specific judicial consideration in Australia. To date, this issue has been more specifically examined at the level of the Patent Office rather than in the Courts. For example, in *Ranks Hovis McDougall Ltd's Application* (1976), which concerned claims directed to a particular fungus, *Fusarium graminearum*, the subject of the patentability of living organisms was first considered. In this case, a patent was granted to a new strain and variants of the fungus. Whilst the organism was itself known, the useful variants or mutants thereof were not and could only be produced via a technical intervention. In considering the case, the Hearing Officer stated:

[16] 18. (1) Subject to subsection (2), a patentable invention is an invention that, so far as claimed in any claim:
- (a) is a manner of manufacture within the meaning of section 6 of the Statute of Monopolies; and
- (b) when compared with the prior art base as it existed before the priority date of that claim:
  - (i) is novel; and
  - (ii) involves an inventive step; and
- (c) is useful; and
- (d) was not secretly used in the patent area before the priority date of that claim by, or on behalf of, or with the authority of, the patentee or nominated person or the patentee's or nominated person's predecessor in title to the invention.

(2) Human beings, and the biological processes for their generation, are not patentable inventions.

> An objection to a claim to a new microorganism, being something living, as not a manner of manufacture, is based, in my opinion, on too restricted a view of the meaning of manufacture in section 6 of the Statute of Monopolies.

Essentially, it was found that merely the fact that the organisms were living was not a barrier to their patentability. In coming to this conclusion, the Hearing Officer set out the following principles:

- New variants require useful improved or altered properties;
- Naturally occurring microorganisms *per se* are not patentable (discovery), but a claim to an isolated culture (on the assumption such a culture is novel and inventive) is patentable; and
- Methods of producing new variants/strains are also patentable, provided the results of the method are reproducible.

In terms of the patentability of higher life forms, such as animals, they are not treated by the Australian Patent Office any differently from the way in which lower life forms are treated, as set out above. Thus, it is the practice of the Australian Patent Office to accept patent applications having claims to transgenic non-human animals, to methods for obtaining animal varieties, methods of cloning animals, and methods of obtaining useful products from transgenic animals (Table 5.1).

Table 5.1. Examples of granted patents of accepted Australian patent applications.

Patent or application no.	Title
654713	Transgenic non-human animal carrying non-infectious HIV genome
685250	Production of fibrinogen in transgenic animals
704486	Animals with targeted gene disruption
716956[a]	Quiescent cell populations for nuclear transfer
737678	Transgenic animal with recombinant vascular endothelial growth factor B (VEGF-B) DNA and uses thereof
741340	Non-human transgenic animal in which the expression of the gene coding for insulin is deleted
742363	Nuclear transfer with differentiated fetal and adult donor cells

[a] Under opposition at the time of writing.

New Zealand

According to Section 2 of the New Zealand Patents Act 1953, an invention is defined by reference to the same term 'manner of manufacture' as is used in the Australian legislation (see above). The practice of the New Zealand Patent Office is therefore similar to that of the Australian Patent Office in granting patents for animal-related inventions. To date, there has been no decision of a New Zealand court concerning the patentability of inventions in this area of technology. See Table 5.2 for examples of patents awarded.

Table 5.2. Examples of granted New Zealand patents for animal-related inventions.

Patent no.	Title
247062	Improved process for bovine nuclear transfer
281041	Transgenic animals producing oligosaccharides and glycoconjugates
332854	Methods for creating transgenic animals
337445	Transgenic animals and cell lines for screening drugs effective for the treatment of Alzheimer's disease

South Africa

Section 25 (4)(b) of the South African Patents Act states that a patent shall not be granted:

> for any variety of animal or plant or any essentially biological process for the production of animals or plants, not being a microbiological process or the product of such a process.

Consequently, animals and animal varieties *per se* are not patentable. At this point, there has been no judicial decision as to what constitutes an 'essentially biological process' or a 'microbiological process'. However, based on the Guidelines for Examiners in the European Patent Office, one might expect that animals and animal varieties, as products of a process, are patentable if such animals are products of a microbiological process. Similarly, processes for the production of animals are patentable only if such a process is not essentially biological or is a microbiological process. In terms of animal materials and processes for the production of animal materials, one would envisage an absence of restrictions on the patentability of such inventions provided their exploitation could not be generally expected to encourage offensive or immoral behaviour.

Japan

In general, patents to animals are allowable in Japan. The patentability of animals in Japan is governed by Part VII of the 'Examination Guidelines for Patent and Utility Model in Japan' (hereinafter the 'Guidelines'), which were first published in February 1997, and revised on 28 December 2000. These guidelines apply to applications filed on or after 1 July 1995.

Chapter 2 of the guidelines is entitled 'Biological Inventions' and governs four types of biological inventions:[17]

(i) those produced by genetic engineering (Section 1),
(ii) microorganisms (Section 2),
(iii) plants (Section 3), and
(iv) animals (Section 4).

The introduction to Section 4 of Chapter 2 provides a non-exhaustive list of 'animal inventions' which are considered patentable, including animals *per se*, parts of animals, and processes for creating and/or using animals.

There are a number of exceptions to patentability with respect to animal inventions. For instance, an invention is not considered patentable if it is merely a discovery, for example a newly discovered animal;[18] if it is incapable of industrial application, such that the utility of the invention is not described, or cannot be inferred;[19] or if it is contrary to public order, morality or public health.[20] Finally, humans are explicitly excluded from the definition of 'animals'.[21]

China

As set out in Article 25 (4) of China's Patent Law,[22] animals, transgenic or otherwise, are not patentable in China. This means that regardless of whether such animals have been produced by traditional biological methods, or whether they have been genetically engineered, they constitute an 'animal variety', and are therefore not patentable inventions (Qingkui, 2000).

However, cell lines, plasmids, nuclear material and 'microorganisms' are patentable,[23] as are animal (and human) tissues which are used for the prevention, diagnosis or treatment of certain infectious diseases.[24] In addition, animal materials and processes for the production of animal materials are also patentable.

[17] Chapter 2 of the Guidelines can be obtained from http://www.jpo.go.jp/index
[18] Section 4.2.1 (1) of the Guidelines.
[19] Section 4.2.1 (2) of the Guidelines.
[20] Section 4.2.2 of the Guidelines.
[21] Introduction to Section 4 of the Guidelines.
[22] The Patent Law of the People's Republic of China came into force on 1 April 1985. Although it has been amended since, notably in 1993 (the provisions prohibiting the patenting of pharmaceutical products and substances were removed), the patentability of 'plant and animal varieties' was not affected.
[23] Rule 25 of the Implementing Regulations of the Patent Law and Chapter 10 of Part II of the Examination Guidelines.
[24] The allowability of 'biological products' is a consequence of the 1993 Patent Law Amendments relating to the allowability of pharmaceutical substances (n. 10). Articles 25 (2) and 25 (3) of the Patent Law, which had previously prohibited the patenting of 'pharmaceutical products and substances obtained by means of a chemical process', were removed, therefore clearing the way for the patenting of biological products.

Brazil

In May 1996, Brazil passed its Industrial Property law[25] in order to satisfy its obligations as a signatory nation of the TRIPs agreement. Under Brazilian Industrial Property law 'living beings, in whole or in part, except transgenic micro-organisms' are not considered to be patentable inventions,[26] whether transgenic or otherwise.[27]

Therefore, animals and parts thereof are not patentable in Brazil. However, processes for the production of animals and indeed animal materials are in principle patentable subject matter.

Argentina

In March 1996, the Argentine executive branch passed legislation on patents and utility models.[28] According to Articles 6 (g) and 7 (b)[29] of this legislation, all classes of living matter pre-existing in nature or genetically modified (with the exception of micro-organisms) are not patentable. However, the Argentine Patent Office does consider inventions relating to animal materials to be capable of patent protection.

---

[25] Law No. 9.279/96.

[26] See Article 10 and Article 18 of Brazilian Industrial Property Law. Article 10, part IX states:

The following shall not be considered inventions or utility models:

IX. all or part of natural living beings and biological materials found in nature or isolated therefrom, including the genome or the germ plasm of any natural living being, and any natural biological process.

Article 18, part III states:

The following shall not be patentable:

III. living beings, in whole or in part, except for transgenic microorganisms meeting the three requirements of patentability – novelty, inventive step and industrial application – in accordance with Article 8 and which are not mere discoveries.

[27] *S*ole paragraph immediately following Article 18:

For the purposes of this Law, transgenic microorganisms mean organisms, except for plants or animals in whole or in part, that due to direct human intervention in their genetic composition express a characteristic that cannot normally be achieved by the species under natural conditions.

[28] Law No. 24.481 on Patents and utility Models (as amended by Law No. 24.572). Its consolidated text was approved by Decree No. 260/96 on 20 March 1996.

[29] Article 6 (g):

The following shall not be considered inventions for the purposes of this Law:

(g) any kind of live material or substances already existing in nature.

Article 7 (b):

The following shall not be patentable:

(b) all biological and genetic material existing in nature or derived therefrom in biological processes associated with animal, plant and human reproduction, including genetic processes applied to the said material that are capable of bringing about the normal, free duplication thereof in the same way as in nature.

Conclusions

The issue of the patentability of higher life forms, including plants and animals, has been an emotive one and will probably continue to be so. In many jurisdictions, statutory provisions have been enacted which outlaw the granting of patents for animals or processes for their production, such provisions being generally perceived by the legislators as justified on moral, ethical or social grounds. Nevertheless, in a significant number of jurisdictions the legislature has not adopted this course and as a result the issue of the patentability of higher life forms has come before judicial or quasi-judicial bodies for consideration.

Increasingly it has been seen that the criteria applied by the judiciary in determining the patentability of higher life forms are no different from those applied when determining the patentability of other forms of innovation. The trend has therefore been towards judicial bodies determining that higher life forms, including animals, are not inherently unpatentable. Active opposition to this position continues to exist, however, and the question of the patentability of animals and processes for their production can probably not be considered to be finally settled in any jurisdiction where no relevant statutory provision exists.

References

2 U.S.P.Q.2d (BNA) 1425 (Bd. Pat. App. & Interferences 1987).

2 U.S.P.Q.2d (BNA) 1427 (Bd. Pat. App. & Interferences 1987).

Application of Abitibi Co. (1982) 62 CPR (2d) 81.

Canadian Biotechnology Advisory Committee, Patenting of Higher Life forms and Related Issues, June 2002. Recommendations 2–5.

Diamond vs. Chakrabarty, 447 US 303 (1980a).

Diamond vs. Chakrabarty, 447 US 303 (1980b), per Burger CJ at 308.

Diamond vs. Chakrabarty, 447 US 303 (1980c), per Burger CJ at 310.

Directive 98/44/EC of the European Parliament on the Legal Protection of Biotechnological Inventions.

EP Application No. 85304490.7, Publication No. 0 169 672A.

LELAND STANFORD/Modified animal (2002a) E.P.O.R. 2.

LELAND STANFORD/Modified animal (2002b) E.P.O.R. 2, page 22, paragraph 47, lines 1–2.

LELAND STANFORD/Modified animal (2002c) E.P.O.R. 2, page 22, paragraph 45, lines 7–8.

LELAND STANFORD/Modified animal (2002d) E.P.O.R. 2, page 23, paragraph 51, lines 1–5.

National Research and Development Corporation vs. Commissioner of Patents (1959a) 102 CLR 252.

National Research and Development Corporation vs. Commissioner of Patents (1959b) 102 CLR 275.

Pioneer Hi-bred vs. Canada (Commissioner of Patents) (1987) 3 FC 8.
Qingkui, Z. (2000) 'Patent Protection for Biological Inventions in China', China Patents & Trademarks, No. 4., 2000: 24–28 at 25.
Ranks Hovis McDougall Ltd's Application (1976) 46 AOJP 3915.
Senate Report No. 1979, 82nd congress, 2nd session, 5 (1952); H. R. Report No. 1923, 82nd congress, 2nd session, 6 (1952).
Straus, J. (1989) The development and status of European law. In: Lesser, W. (ed.) *Animal Patents: The Legal, Economic and Social Issues*. Macmillan Publishers (Stockton Press), New York, p. 21.
T19/90 (OJ 1990, 476), OJ 1992, 588.
World Intellectual Property Organisation (WIPO) (1988). Exclusions from patent protection. Document HL/CE/IV/INF/I.

The Public University, Intellectual Property and Agricultural R&D

6

Thomas A. Fretz[1] and David R. MacKenzie[2]

[1]University of Maryland, Office of the Dean and Director, College of Agriculture and Natural Resources, 1296 Symons Hall, College Park, MD 20742, USA; [2]Northeast Regional Association of Agricultural Experiment Station Directors, Maryland Agricultural Experimental Station, University of Maryland, 120 Hartwick Building, College Park, MD 20742-3131, USA

Introduction

The process of moving a research discovery to the successful adoption of the derived products or services is a necessary function if investments in scientific research are to be justified as a public expense, but the old method of 'open access' technology transfer seems to be failing in some instances. New methods for deploying research discoveries for economic, social, health and environmental benefits are being sought by public institutions. As a result, these evolving policies and changing expectations are clashing on campuses nationwide with more historical professional standards and academic traditions. The sources of these campus tensions are not new, but their intensity is increasing. New management solutions are needed if we are to preserve our most important academic values and traditions, while fulfilling rising internal and external expectations to deploy our research results in ways that will have more of an impact.

Much of the debate over how to manage university research and development (R&D) interests stems from the new applications of molecular biology (or biotechnology) and computer sciences (or information technology). The issue is in deciding how best to deliver university-derived discoveries to the intended public. The considerations are complex, and constantly changing. Traditional technology transfer channels have in some cases proven to be inefficient, some say, while others say we must find new commercial channels to deliver the results of publicly funded research. The core of the tension we face today is found in the difference between the traditions of openly sharing our research discoveries (as 'public goods') and the protection of those intellectual discoveries for commercial purposes (as 'private goods').

This chapter will look at underlying institutional concepts that have created the tensions we face, and the factors impacting the management of publicly held intellectual property, all from the perspective of a public institution. Furthermore, it will look at the consequent trade-offs that must be considered when meeting legal, social and institutional obligations. All of that will be viewed from the perspective of publicly supported institutions that are also charged with responsibility to produce 'public goods'.

Private Goods

Intellectual property consists of personal property resulting from the creative work of the mind or intellect. Intellectual property protection refers to that body of law involving patents (utility patents and plant patents), copyrights, trademarks, trade secrets, plant variety protection certificates, contracts and laws of conversion. Thomas Jefferson's philosophy that 'inventors should receive liberal encouragement' is embodied in the modern system of intellectual property protection. Intellectual property rights are grounded in the US Constitution, which gives Congress the power 'to promote the progress of science and useful arts, by securing for limited times, for authors and inventors, the exclusive right to their respective writings and discoveries'.

In 1790, Congress recognized the need to provide inventors with the protection of the law, and in doing so stimulated the development of inventions by granting inventors exclusive use of their inventions. During this period of exclusivity, the patent owner is protected against competition from others. The first patent act provided protection for 'any new and useful art, machine, manufacture, or composition of matter, or any new and useful improvement thereof'.

A US patent is a legal monopoly granted by the US government to an inventor, which permits the inventor to prohibit anyone else from making, using, or selling his or her invention, usually for (17) 20 years. Thus a patent is a grant issued by the US government giving the patent owner a temporary right to exclude all others from making, using or selling the invention during the term of the patent. As with other forms of property, the right to make, use, or sell a patented invention may be regulated by federal, state or local law.

Copyrights give authors protection against duplication of their work for unlimited periods of time. Copyrights protect the expression of an idea rather than protecting an invention. Most published books, music and movies are today protected by copyright.

Trademarks are used to identify the source of the owners' goods, products or services. If a company's high-value product is associated with a particular trademark, the owner of the product is able to better preserve its market position as a result of brand recognition and brand loyalty, through its legally protected trademark. As an example, Coca Cola's brand name is protected as a trademark.

Trade secrets refer to any information giving its owner an advantage over competitors not having that information. Trade secret law can protect a technology from theft, although the degree of protection afforded is dependent on the technology itself. As an example, Coca Cola's drink formula is protected as a trade secret.

Patents are not intended to protect secrets or withhold technical information from the public. Quite to the contrary, a valid patent must provide sufficient information so that 'one of reasonable skill' can duplicate the invention described in the patent. This 'enabling information' is disclosed so that others can learn from the invention, improve upon it and develop new inventions, thereby advancing technology. Owners of intellectual property (IP) can either sell or lease their IP, or enter into contractual and licence agreements with others, regarding the property's use.

Two particular parts within the current body of patent law speak to agricultural issues. The Plant Patent Act (PPA) of 1930 granted plant breeders an exclusive right to propagate new varieties by asexual methods, and was designed to encourage research investment in asexually reproduced plant species. PPA is administered by the US Patent and Trademark Office (PTO).

Plant variety protection with certificates was promulgated in the Plant Variety Protection Act (PVPA) of 1970 to provide patent-like protection to sexually reproduced materials, since protection was not available under the PPA. The PVPA, interestingly, is administered by the US Department of Agriculture (USDA), not the PTO. The PVPA was amended in 1994 to change eligibility from 'first to discover' to 'first to file a claim', brings the US more into agreement with much of the rest of the world. The amendments of 1994 to the PVPA also added coverage to tuber-produced crops (e.g. potatoes). Tuber crops had been excluded from the original Plant Patent Act, as a food. Finally, the 1994 changes clarified the rights to saving seed for replanting, and the concept of 'essentially derived varieties' to define better the distinctions required to claim a new variety.

In the early 1980s, US case law and administrative decisions extended common utility patent protections to all forms of living organisms. This too has had an enormous impact on the opportunities and responsibilities of public institutions to protect their IP, especially in agricultural research. The patenting of life was first permitted in the *Diamond vs Chakrabarty* court decision that allowed a utility patent claim on a microorganism. This was followed by *Ex Parte* Hibberd, that allowed a patent on a plant (maize). Recently (December 2001), the US Supreme Court affirmed these case rulings and administrative decisions in a majority opinion written by Judge Clarence Thomas.

Public Goods

Public goods (e.g. products, services and knowledge) are defined as freely available to all and not diminished by use. Illustrative examples of public goods are: food safety guidelines; best farming practices; and, integrated pest-management methods.

For the most part public institutions are responsible for providing public goods, because such socially necessary research investments are unlikely to yield private companies sufficient return (or profit) on their investments. This is to some, a public research and education institution's *raison d'être*.

However, public institutions may also produce private goods, or goods and services that could have commercial value. Illustrative examples are: computer software and databases, improved varieties of crops and animal vaccines, to name but a few. Public institutions are then put in the position of deciding how best to commercialize those private goods, if at all.

There are basically two strategies for protecting the IP of a public institution. Some university-derived IP is placed in the public domain (a.k.a. a 'public trust patent') by claiming IP protection with no intention of going beyond making sure that everyone has open access to the discovery. The second strategy is to seek IP protection with the intention of commercializing the creation, discovery or invention, either directly or through limited licensing.

The first strategy is important for preventing others from claiming someone else's discovery as their own, and thus foreclosing the public's open access to that IP. Fortunately, in the US, the 'first-to-discover' doctrine can also be used to stop others from appropriating property intended for open public use. It is fairly easy to demonstrate that 'prior art' contradicts a first-to-discover claim, invalidating any patents that might have been wrongfully issued, and recording one's discoveries in scientific journals is an effective means for establishing 'prior art'. Unfortunately, only the US and The Philippines employ the first-to-discover doctrine when deciding the merits of a patent application. The rest of the world uses the 'first-to-file' doctrine, allowing for a lot of IP misappropriations, unless an actual patent is issued that is to be held in 'public trust', and that can get expensive.

The second strategy is important for preventing 'generic' offerings by others, especially in those cases when commercialization costs may be substantial and need to be recovered through direct sales or licensing royalties, or through partnering with private interests.

The Role of the Public University

Universities play an important role in the development of science and technology. As centres of higher education, universities are involved in educational and research activities that have served as significant sources of knowledge for society. Universities have determined that their role in society is largely focused around a three-part mission that includes discovery, outreach and instruction. The fulfilment of the threefold mission places the university in a privileged place as a generator of new knowledge.

Public agricultural research has been conducted for over a century and is the primary responsibility of the federal-state partnership supported Land Grant University system and its affiliated State Agricultural Experiment Stations (SAESs), working in collaboration with the USDA. The success this system has generated has benefited from a focused programme with local and/or regional inputs. In addition, the system has had excellent cooperation with other public agencies, as well as private sector. Much of this success can be attributed to the free exchange of information, and the development of a cooperative infrastructure that allows for new information to be acquired, maintained, shared, developed, enhanced and ultimately put to good use, but that was all created before the emergence of IP protection issues.

The Land Grant University system has a historical tradition, moral obligation and a legal responsibility to transfer new inventions, discoveries and technologies to the intended stakeholders.[1] This responsibility now includes, some have argued, an obligation to see that, when appropriate, the products of research reach adoption through commercialization, and thus appropriate commercialization mechanisms are needed to help insure technology transfer. The question then becomes, if we accept this premise, not whether we transfer any of our technologies commercially, but how and when we transfer technologies, and what role we as faculty and administrators of those research programmes have in these decisions?

---

[1] Land Grant University faculty and research administrators should periodically review the language of the Morrill Land Grant Act of 1862 and the Hatch Act of 1887. There one will find that there is an inherent obligation to place their research results before the public for adoption and use.

The Dilemma

Traditionally, universities have been excellent sources of knowledge for the private sector, and at the same time the private sector has benefited from access to newly educated professionals. More recently there has been an incremental change in the relationship between the academic and private sector. Many universities are now opening to the challenge of getting involved with the private sector in a more systematic and managed manner.

The diminished financial environment within which the SAESs operate within today's Land Grant Universities is driving decision making. Without question, both university and SAES research budgets are constrained to the point where they can support only the bare minimum required for high-cost, basic and fundamental research programmes. Moreover, the cost of research is escalating rapidly and thus there are increased pressures to expand the funding base. Universities are hiring more science discipline-oriented scientists, and who in turn rely more heavily on competitive grants and contracts (both public and private) for their support. This has occurred as state contributions each year make up a smaller portion of the institution's budget. Stronger university–private sector relationships and greater dependence on revenues generated through the management of IP are now seen as potential, and possibly significant, sources of research support.

Data from various governmental agencies indicate that the number of patents being issued to universities is increasing significantly. In fact, more universities are increasingly using the number of patents issued as a measure of research success, replacing the published peer-reviewed journal article as a prime metric for faculty evaluation.[2]

The US issues more patents and publishes more scientific papers than its trading partners, yet we lag significantly in transferring new technologies to the marketplace. Perhaps as a response to this disappointment there are a growing number of public and private universities that have established corporate research parks. These investments are intended to stimulate the development of high-tech industries as a means of encouraging local and /or regional economic development.[3]

Increasingly, state governments are establishing technology centres or development agencies to stimulate 'marriages' between a research university and high-tech industries, often with the use of seed money to accelerate the effort. Whether or not these efforts will stimulate basic and fundamental research is yet to be seen.

---

[2] Research administrators should be very careful in using the number of patents issued as a measure of productivity, for it sends mixed signals to the faculty.

[3] Interestingly, this also sends a mixed message to faculty regarding expectations for productivity and advancement, quite possibly at the expense of serving the public's need and wellbeing.

With the widespread recognition of the growing importance of IP management, new university management positions that specialize in 'technology transfer' have recently emerged on most campuses. The initial impetus for creating a technology transfer office can be attributed to federal legislation designed to promote technology transfer. Most notable is the Bayh–Dole Act of 1980, which obligates universities to commercialize the results from the creations, discoveries and inventions resulting from federally funded research. However, beyond requiring the protection of IP and expecting its commercialization, the Bayh–Dole Act allows the collection of royalties and fees that can provide needed income streams to universities, to enhance further their research programmes.

This 'sea change' in IP philosophy has caused considerable consternation on some campuses, and outright strife on others. This is because many within our university communities have come to view the transfer of new technologies as simply 'business ventures'. We suggest, however, that this is a misplaced vision and an inverted priority. It is clear that the greater purpose derived from our Land Grant University mission is to serve the agricultural and food system communities so that it remains viable, competitive and profitable. This distinction makes the Land Grant University community somewhat unique, and amply justifies the continuation, as well as justifiable increases, of funding for the federal–state partnership in agricultural research and extension.

The Trade-offs

The opportunity to manage IP resulting from the work of agricultural scientists has opened many opportunities for research cooperation between universities and the private sector, but this opportunity comes at a price. For each of the claimed benefits, opponents will cite some drawbacks. Listed below are some of the current 'pros' and 'cons' to these issues, hopefully without the subjective judgement of the authors.

Establishing trust

Pro: Research partners can more confidently approach the other with less risk, knowing that ideas and valuable technical information will be protected and not compromised if there is a formal research partnership.

Con: Selecting one research partner over others creates problems for a public institution, and adds to distrust within the broader stakeholder community.

Access to findings

Pro: Private sector partnering with public universities gains through early access to findings from basic research programmes. The private sector also gains access to potential employees. Some companies, especially smaller and/or start-up firms also can gain access to special research services.
Con: Exclusivity in the access to research findings is contrary to the fundamental principles upon which a public university rests. A tax-supported institution cannot play favourites with selected research partners.

Increase in research funding

Pro: Limited funds for agricultural research, especially in the basic sciences, especially when applied toward the agriculture and food system, have resulted in the neglect of many important problem areas. In working with universities, the private sector is genuinely interested in basic research, research that cannot or will not be conducted as readily, or that has limited potential for commercialization.
Con: Private partners in a public university research project rarely pay the full cost of discovery research. Most pay only a few cents on the actual dollar cost. Why should they have an insider's access to research results that others have paid for with tax dollars? Obviously, they should not.

Focus on relevant issues and problems

Pro: Private sector involvement in university research helps to insure that research funds are focused on those problems that have potential for pay-off, and address relevant issues. Such focus is consistent with the goal of maintaining agricultural/life science and private sector competitiveness.
Con: By allowing private partners to set the research agenda, other stakeholders with different needs are excluded from participation and the benefits. This is an untenable position for a public university.

Improved technology transfer

Pro: In some cases, the results of university research fail to capture the attention of the private sector and investors, as they are unable to translate findings into products and services. Private sector involvement in the research programme can help achieve effective translation of research results to products and services.

Con: Recognizing the appropriate use of private interests in transferring technology has challenged universities from the beginning. Most institutions end up selecting those research topics that seem to be most likely to yield big royalties, and pass on the most socially relevant or environmentally significant opportunities. Until there is a better system for distinguishing the important from the opportunistic, this argument is fraught with problems.

Employment opportunities for students

Pro: A significant and growing portion of the graduate students trained in university research programmes find employment with the private sector. The opportunity to interact with scientists from the private sector during the research programme improves the students understanding of the corporate research environment and establishes valuable contacts upon entry into the job market.
Con: Many students are reluctant to tie their research projects in with private interests as publication of their results (a must for gainful employment) may get embargoed for commercial reasons.

Ready access to research findings

Pro: Universities and federal laboratories are recognized as primary sources of basic research. When technology is evolving rapidly, access by the private sector to the findings of basic research programmes is valuable. Sponsorship of research provides an opportunity for early access, not only to protect the IP, but more importantly it provides the private sector access to valuable know-how. This know-how might include the techniques, methods and materials usually associated with a state-of-the-art research programme.
Con: For many emerging technologies (such as biotechnology) this has not proven to be true. Many of the large companies have more advanced (and proprietary) research knowledge, making access to public universities irrelevant. Most of the more campus-active corporations may be interested in 'fishing expeditions', but not for active partnering.

Contact with leading researchers

Pro: Through research partnerships university and industry researchers can establish contact with experts in the reciprocal sector that might be difficult to establish otherwise. In some situations these contacts can be important and productive consultancies.

Con: The possibilities for conflicts of interest are numerous when public–private research partnerships form. This is especially true when consultant fees are paid for services, and, some universities are notoriously poor at policing this aspect of its 'outreach'.

Student contacts

Pro: Companies look to universities as the primary source for technically trained employees. Sponsored research relationships offer the private sector the opportunity to become acquainted with students and post-doctoral scientists. Additionally, students are usually not constrained by the experiences and prejudices as one might find with well-established investigators, hence, students are sources of imaginative new approaches to researchable problems.
Con: The primary mission of a university is to provide education to students. That experience needs to be carefully managed. Contacts with private firms may, in some cases, not be in the student's best interest. The selection of research topics, the enforced timing of publications (noted above), and the quality of student scholarship may be called into question when there is some private interest involvement. Too few institutions have mechanisms in place to guard against such academic interference.

Access to special research services

Pro: Within the private sector, smaller and/or start-up firms often lack the capacity for special research services, such as amino acid analysis or DNA sequencing. Collaborations with the university research community can make such services available.
Con: Sharing publicly funded facilities with private interests, especially if done exclusively, is contrary to the best interests of a public institution, and can only lead to conflict.

Access to IP

Pro: Most private sector–university collaborations provide the private partner with some form of access to their IP. Admittedly, there is a risk that some in the private sector might benefit disproportionately from such relationships, but the benefits of contributing to the private sector through licensing technologies developed collaboratively can be substantial.
Con: Such private access to publicly funded research results is inappropriate for a public university in that the licensees may acquire exclusive or limited exclusive access to technology developed at least partially with public funds.

Implications for Society

As some have argued, providing the private sector with access to technology acquired through research partnerships is inappropriate, in that the private sector may acquire exclusive access to technology developed with public funds. This position, however, must be weighed against the prospect that, without an opportunity for the private sector contributors to license the technology, and to thus recover development investments, new technologies may go undeveloped. If this should happen, no one benefits from the public's research investment, and society foregoes the economic gain that might be brought about by new business ventures and new value-adding technologies.

There is however, no reason for this to happen. Measures such as limitations on the period of exclusivity; reasonable royalty rates and/or fees; and, march-in rights[4], can all help to assure that the public's best interest is being served.

The response

How do we as a community of scholars respond to these issues and challenges? Should we cling to our traditional approaches, putting all discoveries on the public shelf, available to everyone? Or should we be moving to a new philosophical position on the management of IP?

University-developed technologies can reduce agricultural production cost, provide for a more sustainable agriculture, address emerging environmental issues, increase farm profits, and even provide the general public with lower cost, safer, more nutritious and healthier food. To see those technologies fail to be adopted because of the lack of an incentive for commercial development seems wasteful. Thus, some degree of partnering with the private sector, and the licensing of IP to private interests is necessary, but this is not sufficient. Some careful thought is needed as to when and under which circumstances such public–private partnerships are the best solution for transferring technologies. Opening up the university research system to private interests may be critical for letting the marketplace determine the real value of the property, but the public institutions' purpose in each case must be to consider the need to advance the technologies that they have developed, either by the traditional public channels or through private commercialization. The key to getting this right is in the decision-making process through which institutions reach their final choice regarding

[4] Provisions that allow the owner of a technology to reclaim the property from the licensed company if it fails to fulfil its obligation to develop and market the technology under the terms of the agreement.

the appropriate means for transfer of technology. We believe that this needs to be done on a case-by-case basis, and with the corresponding research faculty's input. Importantly, many factors need to be considered in each case.

The research community has, to a great degree, foregone this obligation and forfeited much of their responsibility regarding IP rights (IPR) management to their central university's decision makers, resulting in research laboratories being seen as *'profit centres'*, useful for replacing public funding that has been lost over the past two decades. We recognize that universities need to meet budget shortfalls in the public revenues used to support needed agricultural and biological research, but it is too easy for universities to rationalize that, with diminishing appropriated funding, they need to license all discoveries to the highest bidder. Public institutions have an obligation to transfer the results of their research to the general public, as a returned payment for their continuing tax-based support. There are circumstances that warrant the use of patent licensing or other forms of IP protection to insure that a discovery is commercialized, but that is not always the case.

Faculty–inventor expectations

From the public university's perspective and that of research administrators, there are issues from the point of view of the inventors, owners and users of IP. What do the inventors and owners expect to accomplish when seeking IP protection of improved germ plasm, bioprocesses, analytical techniques, etc.? Why do they seek protection? Do the motivations and the benefits expected from IP protection match with the mission of the institution? What are the responsibilities and expectations of potential users of protected materials? What latitude should be allowed for the use of protected materials for further research purposes? What does the public institution expect with regard to benefits and expectations of IP protection?

Recognition of research accomplishments and accountability are extremely important in today's world. IP protection, regardless of the type, is a form of recognition and reward for conducting quality, meaningful research. This recognition is important for faculty morale, and in many cases to the tenure, promotion and merit review processes as well. IP also provides a form of protection from economic interests. In some instances, this may be with the intent of developing and/or securing funding sources needed to keep research programmes viable.

In short, what do scientists expect to accomplish and why do they seek IP protection for their work?

Stakeholder interests

One of the fundamental challenges facing the publicly supported university today is to develop effective strategies and mechanisms for educating the public and their elected officials on the economic, social, health and environmental benefits of basic and applied research in agricultural and the life sciences. There is a real need to support the range of basic and mission-linked research as an investment in the long-term cumulative process that generates solutions to practical problems. The public needs to know that, if public tax support is to continue, relevant, responsive and high-quality scientific research will result.

Some public interest groups, however, seem to perceive that some part of today's public agricultural research effort is unjustified, and should be conducted in the private sector without the benefit of public funding. As the private sector begins to shoulder a larger part of the responsibility for applied agricultural research, it will become increasingly difficult for the general public to accept the merits of publicly funded agricultural research.

Continued erosion of state and federal funding for research has generated pressure on university researchers to seek funds from other sources. As noted above, other funding sources may include the private sector, from which funds can normally be secured much more readily if universities are willing to protect IP and grant licences to sponsors of such research. As a consequence of such arrangements, the goals of long-term university-funded research programmes may be redirected by the provider of the marginal funds. It is this potential for the redirection of programmes by the marginal provider of funds that has university and agricultural research administrators concerned.

Redirection of the research agenda may be more directly related to funding sources than intellectual property protection, *per se*, but it is present regardless of whether the marginal funds are provided by the private sector or by a federal granting agency. The provider of the marginal funds can have an enormous and significant impact on the direction of the research programme. Additionally, the prospect of collecting royalty income from the licensing of IP may also affect the research priorities of some institutions and for some individual researchers.

Decision makers and researchers often lack sufficient information and background about IP Rights and appropriate institutional procedures or policies for exercising these rights without compromising public interest. Decision makers should be better informed of the impact of IP protection shifts on the direction of research programmes and the impact of such directional shifts on the ability of these programmes to meet the public's expectations.

Principles of IP Management

Arguably, while this is an emerging field of research management, we offer, with hesitation, some guiding principles that can help to focus the issues surrounding the management of IP generated by publicly supported universities.

Principle 1

It is clear that the marketing of IP is a legitimate activity of a university, but the faculty and administration must understand that, in addition to the income that is placed back into research programmes, the intended technologies must serve a public need.

Principle 2

It is critical that as university efforts to commercialize discoveries increase, faculty and research administrators must not allow that activity to give priority to short-term, low-quality research to boost an institution's or an individual's financial gain.

Principle 3

The decision on how to handle and manage the IP developed by the faculty must always be balanced against the substantial costs required to protect and market inventions, versus the need to have the technologies made available to those who have invested in the technology, most notably the tax-paying public. This is not an easy balancing act.

Principle 4

Within the academic community there must be due diligence when collaborating with the private sector, as a university must not sacrifice its independence or its integrity.

Principle 5

It is important that each academic institution clearly understands and pursues its mission, and, in turn, exercises care in its relationship with the private sector.

Principle 6

It is important for any university to strive to diminish any potential conflicts that might be derived from its management of IP.

The above thoughts lead to several frequently asked questions regarding IPRs and property protection generated by university scientists. We offer our perspectives as answers.

- Who should be involved in the decision-making process for protecting, licensing and managing the academic institution's IP?

This should primarily be done by the involved research faculty, with the proviso that they are informed on the issues of *IP* management policy.

- Who should assign the IP rights to the creations, discoveries and inventions of an academic institution?

This should primarily be decided at the faculty and department level with concurrence from the college administration. Otherwise, too many conflicting interests get involved.

- What mechanisms are available to assist faculty to become knowledgeable on the issues of IP?

Few mechanisms if any exist at this time. The university administrative community should work collectively to fill this gap as soon as possible.

- What is the role of the university, department and faculty in the commercialization of new products, technologies or processes that will benefit the agricultural community and the general public?

While this is not true of all institutions, departments and faculty who are impacted by such decisions should be front and centre in both the discussion and decision-making process involved in commercialization of IP. Currently, this is not the case in many institutions.

- How should the process of identification of the IP and the technology transfer activity of the university be structured or otherwise institutionalized?

In the best of all worlds, each academic department should develop its own policy on IP management; however, owing to the difficulty of such, and the potential outcome that the university might end up with dozens of IP management schemes, it would be wise for the university to have guidelines that focus the process.

A policy statement should be developed to take into account the traditions and values of that department/discipline, the mission they are committed to, and the realities of the types of research they are engaged in, and plan to be engaged in. In reality, the most common approach in most universities is to defaulting to the central administration; however, faculty must find models that improve the level of engagement on these matters.

- Who should provide general guidance to those who will make the final decision on which property or properties to protect and by what mechanism?

The existing 'Technology Transfer Offices' should be transformed into a service function for the academic departments. No longer should they drive the IP agenda for the faculty. The faculty should work under broad institutional guidelines and clearly stated principles, with facilitation from centralized support offices.

- How do we (or should we) license all discoveries and IP that result from publicly funded research programmes?

Not all discoveries should be (or indeed need to be) protected, and not all protected IP needs to be licensed. Each case is different. It is the *intention* that is important to the decision, and faculty involvement in deciding on a case-by-case basis is deemed the best possible configuration for a publicly supported institution.

Recommendations

Given the summarized views, the derived guiding principles and our perspectives as research managers, we offer the following recommendations:

- *We recommend* that the faculty–inventors who are/were involved in the discovery research must participate in the decision-making process for the transfer of that technology, and in turn, the involved faculty has an obligation to become informed on the issues related to IP management and its commercialization.
- *We further recommend* that the final decision on technology-transfer strategies must reside with the faculty and their department, and not with the central administration. In the end it is the faculty, the department and the college administration that have the greatest appreciation for the political and societal consequences of a decision to license IP, particularly when public (state and federal) funding has been used extensively.
- *We recommend* that faculty weigh their need to generate funds to support their continuing research in the face of declining funds, against the likely societal or economic impacts associated with each option for technology transfer, and against the political consequences of protection and licensure of property. We note that the consequences of a wrong choice can lead to a loss of years of goodwill and trust from those who have supported agricultural research.

- *We recommend* that training and information on IPR issues be made more widely available to the faculty, especially within the agricultural and life science research system. Additionally, the community of faculty needs to take a genuine interest in understanding the issues and consequences of decisions made that might impact their future.
- *We recommend* that effective efforts be given to counteract the serious erosion of public financial support for agricultural research. Better public funding would mitigate many of the negative impacts of IPR protection on public research agendas. This would occur by removing much of the pressure to acquire from the private sector marginal research money where IP rights must necessarily be an important part of the funding arrangements.
- *We recommend* that public institutions improve their capacity to appropriately manage IP in a way that will encourage rapid commercial development of research findings, while the same time protecting the public's interest. This might include segmenting budget lines to better track public versus private investments, more transparency in the management of royalty streams derived from licensing, and greater involvement of all stakeholders in the development of IP management policy.

References

Caldwell, B.E. and Schillinger, J.A. (1989) *Intellectual Property Rights Associated with Plants*. ASA Special Publication Number 52, Madison, Wisconsin.

Baenzinger, P.S., Kleese, R.A. and Barnes, R.F. (1993) *Intellectual Property Rights: Protection of Plant Materials*. CSSA Special Publication Number 21, Madison, Wisconsin.

Eberhart, S.A., Shands, H.L., Collins, W. and Lower, R.L. (1998) *Intellectual Property Rights III. Global Genetic Resources: Access and Property Rights*. CSSA Miscellaneous Publication, Madison, Wisconsin.

Fuglie, K.O. and Schimmelpfennig D.E. (2000). *Public–Private Collaborations in Agricultural Research: New Institutional Arrangements and Economic Implications*. Iowa State University Press, Ames, Iowa.

Santaniello, V., Evenson, R.E., Zilberman, D. and Carlson, G.A. (2000) *Agriculture and Intellectual Property Rights: Economic, Institutional and Implementation Issues in Biotechnology*. CAB International, Wallingford, UK.

NIMH Technology Transfer Legislation Summary. <http://intramural.nih.gov/techtran/legislation.htm>

7

Lessons from the Patenting of Plants

William Lesser[1] and Martha Mutschler[2]

[1]*Department of Applied Economics and Management, Cornell University, 405 Warren Hall, Ithaca, NY 14853-7801, USA;* [2]*Department of Plant Breeding, 303 Bradfield Hall, Ithaca, NY 14853, USA*

Introduction

Plants of various forms have been protectable with intellectual property rights (IPRs) since 1930, with the specific forms of Plant Breeders' Rights (PBRs) formalized in 1961 and patents first granted in 1985 (see Chapter 1). This significant historical period provides a basis on which the implications of animal patents can be appraised. The assessment is in part limited to agriculture (including crops and ornamentals), for the existing protection systems are largely restricted to cultivated plants. Given the broadly similar structure and economic forces on crop and animal agriculture, some degree of projection is possible. Other categories of patented animals, particularly the dominant one of laboratory models, are utilized in a wholly different institutional system, for which examples from agriculture provide fewer but none the less relevant insights.

The general roles of IPRs, and patents in particular, are the same whatever the area of application. These are the provision of an incentive for investment in new strains and the facilitation of transfer, particularly to other countries where patents are also available and held (see Chapter 1). The purpose of this chapter is the review of experiences under IPR protection for plants, and projection of those experiences for patented animals. Principal emphasis is on experiences in the USA, given the dominance of protection mechanisms there, but experiences in other countries, including developing countries, are included when available.

IPR Protection of Plants in the USA

Under US law, plants are protectable under three separate laws, as follows:

- asexually propagated, Plant Patent Act of 1930,
- sexually propagated, Plant Variety Protection Act (PVPA) of 1970 (hybrids since 1994 only), and
- all plants by patents, *Ex Parte Hibberd* (1985).

As is explained in detail in Chapter 1, the scope of protection for plants using utility patents is substantially broader than that under Plant Variety Protection (PVP), because patent claims can include restrictions against saving seed from the crop as a subsequent seed source or use of the patented materials by other breeders in a breeding programme without permission. (PVP may alternatively be called Plant Breeders Rights, or PBR.) Patents for plants (excluding the specialized Plant Patents) have been available only since 1985. The effect of these several factors on protection decisions by seed companies can be seen by examining the records of applications for PVP and patents on seed-produced crops.

PVP applications for all crops/species

The entire data set of applications for PVP to the US PVP Office was downloaded for analysis[1]. Since PVP applications are dated upon receipt and rapidly added to the database, rather than being confidential until issuance, this data set is complete through 2001. To analyse the use of PVP to protect intellectual property (IP), the changes in applications for PVP over time were examined. Application dates rather than grant date were used, since application is a good indicator of interest in protecting plant IP and is not affected by the processing lag or other external factors.

Trends for total annual applications were first considered for all crops from 1971 to 2000 (Fig. 7.1), during which period there was a general rise in applications. The rise though is not uniform as a whole or in major species/crops. Dividing the data into quarters instead of years, or dividing the data by applicant as well as time, did not reveal any particular patterns.

A jump in the total number of applications occurred in the early 1980s, coinciding with legislative changes adding several previously excluded vegetable crops to the list of protectable species, but this aberration was limited and short-lived; the involved crops are of minor economic importance. Also observable is an increase in application numbers

[1] http://www.ams.usda.gov:80/science/PVPO/pvp.htm

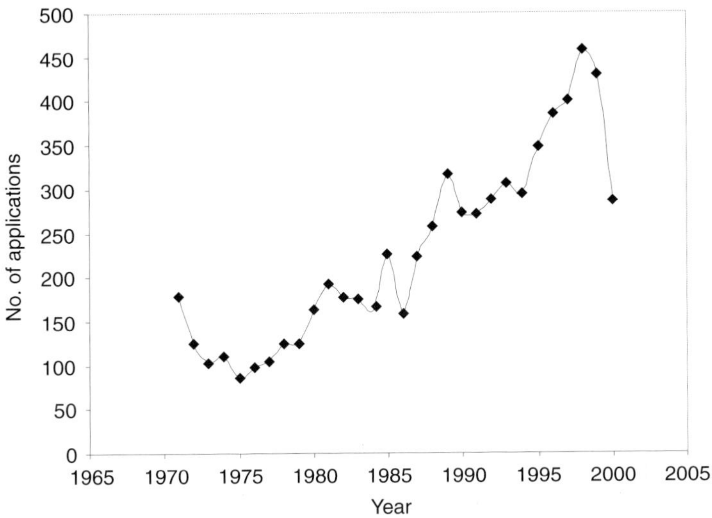

Fig. 7.1. Annual applications for crop patents to US Patent Office.

around the mid-1980s continuing throughout the 1990s. However, the most marked change is the substantial drop in applications in 1999 and 2000. The PVP Office confirmed that this is a real decline in applications received, and not a function of when and how data are added to the database.

Partitioning total application numbers into crop-specific values reveals underlying trends, in particular the significant differences in applications across crops. Total applications since 1970 include 189 crops/ornamentals, with a third of the crops represented by only one application and the top two crops (soybeans and maize) accounting for a third of all applications. Indeed, the top five crops (soybeans, at 19%, maize at 12%, bread wheat at 7.7%, cotton at 6.1% and lettuce at 5.4%) combined, account for 50% of all applications. Reasons why a few crops might account for such a large part of the total PVP applications are associated with economic value and mode of reproduction.

Wheat, soybeans and cotton are the major US non-hybrid crops in terms of acreage planted and crop value. They are also all self-pollinating crops, and as such are readily maintained as stable uniform lines. Hence, growers can and do save part of the crop as a seed source rather than purchase seed annually. Due to the reproductive biology of these crop species, control through the sale of F1 hybrids is not economical.

Maize is also a valuable crop, but is different from wheat, soybean and cotton in that field maize cultivars are almost invariably hybrids. However, the parent inbred lines are uniform and open-pollinated and so could be protected by PVP, even before the change of the PVP statute in the early 1990s extended protection to F1 hybrids. Commercial maize inbred lines are extremely valuable. Companies take great care to prevent the release of pure lines, but it is impossible to exclude totally seed of the inbred lines, particularly the female line, from a hybrid seed lot. This provided the impetus for considering legal means of protecting inbred lines and the hybrids they produce.

Some vegetable crops do appear within the top 10 PVP protected species. Lettuce is usually the second most valuable vegetable crop in the USA, after tomato. Lettuce is an out-crossing species, but does not have an economically viable means of producing hybrid seed due to its flower structure and reproductive biology. Therefore, lettuce varieties are largely self- or open-pollinated. In contrast, tomato, which is usually the most valuable fruit in production, is self-pollinating, largely represented in cultivation as hybrid varieties, and accounts for less than 20% of the PVP applications of lettuce.

Examining changes in applications for the leading protected crops and PVP applications as a whole, it is apparent that the increase in total applications is largely attributable to the increase in soybean applications, and subsequently to the protection of field maize inbred lines. Soybean PVP applications rose throughout both the 1980s and 1990s. Applications for PVP on field maize pure lines started in the mid-1980s, despite the fact that inbred lines are not commercial products themselves, and increased rapidly thereafter through the 1990s. As a result of these changes, field maize and soybean applications rose to nearly 45% of total applications in 1998. If it were not for this major shift in these two crops, the total number of PVP applications would have experienced only a slow increase through the mid-1990s.

The impetus for maize PVP applications occurred during the prosecution of *Pioneer vs. Holden* (1987), with applications for utility patents on maize lines and hybrids following soon thereafter. That case, ultimately won by Pioneer, concerned misappropriation of maize inbred IP. Even though the use of trade secrets was upheld in that instance, breeding firms recognized the limitations to relying on trade secret legislation exclusively for protecting valuable breeding lines, and turned to patents and PVP.[2] The increase in PVP applications noted in the 1990s coincides with the drafting and implementation of changes in the PVP statutes, broadening the protection under PVP to include F1 hybrids (often produced with lines protected by a PVP; see Chapter 1).

The peak and then decline in PVP applications in the late 1990s is puzzling as it is in counter-distinction to the decades-long trend of increased use of IPRs. Several possible explanations exist, although a few additional years must pass before the existence of a trend can be identified. Among the possible explanations are the delays by the PVP

[2] In simple terms, a small (< 1%) of the female pure seed will be produced and intermixed with the hybrid. This is the seed the court case involved. The male line may also appear, but very rarely.

Office in processing applications, although a level of interim protection is afforded by PVP applications in process. Another possibility is the rise in the use of transgenic crops; by the 2000 crop year about 55% of soybeans and 25% of maize were transgenic varieties (www.usda.gov under ers). As PVP do not provide complete protection for transgenic crops (see Chapter 1), the use of patents has been sought as an alternative. Finally, it has become evident the US version of PVP provides only minimal protection from near copies, primarily because the degree of 'distinctness' required (see Chapter 1) is very limited. Indeed, the PVP Office generally accepts any claim, no matter how minor or impractical the claimed attribute. Hence, competing firms sell essentially the same genetic material so that Lesser (1987) has referred to the US PVPA as effectively protecting the variety name rather than the genetic material. He (Lesser, 1994) has also measured the value of PVP for soybeans in New York State to be a modest 32 cents on a $14 bag of seed. As a consequence, firms may be selecting patents over PVP as a stronger form of protection.

Patents for maize and soybeans

Patent numbers for maize and soybeans are shown in Fig. 7.2 for the period 1985 (when first allowed) until 2001. All other crops have registered only a handful of applications up to the present. Because the US Patent and Trademark Office (PTO), until spring 2001, did not publish applications unless and until a patent issued, the only detailed data are on issued patents, as shown in Fig. 7.2. Because the examination period takes an overall average of about 2 years, but can take longer and occasionally much longer, it is not possible to know at the time of writing if the patent grants shown for 2001 under-represent the actual values.

The data are from the US PTO web site searchable database,[3] where it is possible to identify patents by class or by a key word. For the data used in Fig. 7.2, it was found that maize patents for example were filed under a number of key words, including corn, maize, and/or the scientific names *Zea mays*. Titles and abstracts were analysed to exclude patents on items outside of the focus, including but not limited to agricultural equipment, sweet or popcorn, creation of transgenic plants, and process patents (such as breeding techniques for maize). A line for applied use that also happened to be transgenic would be retained, the transgene being considered the same as any other gene in the lines' genome. Hence, the figure represents a careful compendium of traditionally bred maize and soybean patent numbers by date of application.

---

[3] www.uspto.gov

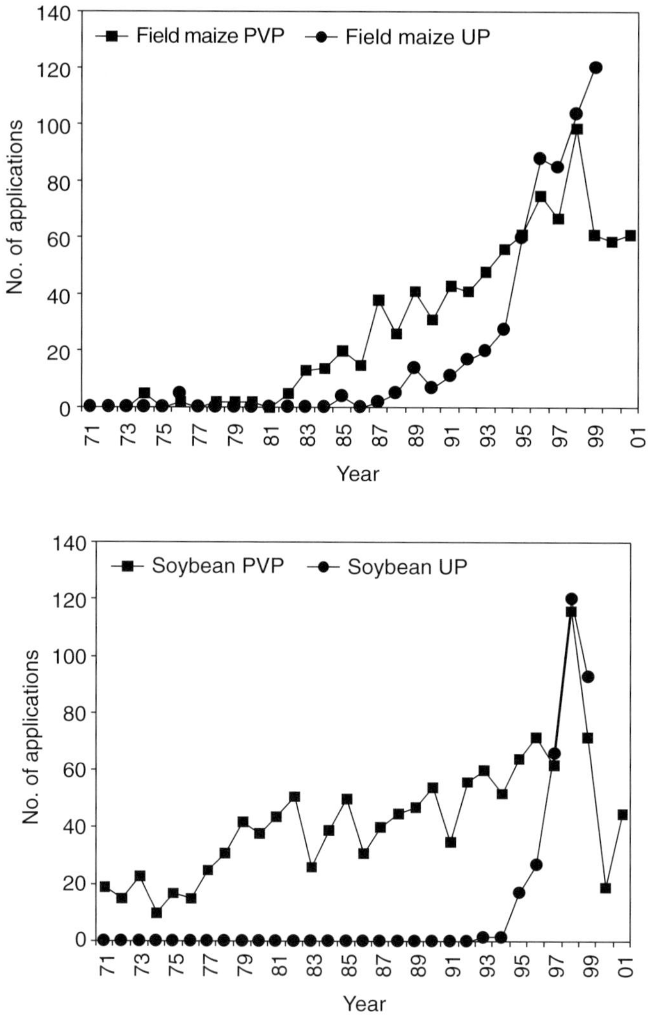

Fig. 7.2. Number of patent applications for field maize and soybeans.

Examining the figure indicates several apparent flex points in application numbers, particularly for maize. One occurs around 1987, and coincides closely with the *Pioneer vs. Holden* decision noted above. Pure line seed is protectable with PVP, but the existence of PBRs makes enforcement complex so that firms appeared to opt for patenting (Seay, 1993). About 1992, one firm, Pioneer Hi-bred, is believed to have begun protecting hybrid seeds. At the time, US PVP legislation did not allow for the protection of F1 hybrids, leaving patents as the sole vehicle. That restriction was ended in 1994, but firms continued to seek patents anyway.

A final jump in patenting of maize is seen around 1996. That year also marked the first for the commercial introduction of transgenic crops in the USA. The new product introduction likely explains much of the increase in patents for both maize and soybeans following 1996.

Patents for crops other than maize and soybeans

In contrast to the situation in maize and soybeans, there are very few utility patents for lines, varieties or hybrids of other agronomic or horticultural crops. This is not due to a lack of interest in protecting materials, as these crops are well represented in PVP applications. Nor is it due to a lack of awareness of the potential of patents for protecting materials. Rather, economics is the likely explanation; returns on seed of even the top varieties of non-maize and soybean crops are small in comparison so that the seed industry is less able to support the costs of obtaining and enforcing patent protection.

Fairly basic economic factors seem to explain much of the application of both PBRs and patents for plants by seed companies. The use of PBRs rose rapidly following the 1970 introduction, but only for certain crops and not others. Foster and Perrin (1991) found a strong statistical relationship between the value of a crop and the number of PVP certificates granted. As the data in Fig. 7.1 show, that means that there are many certificates for soybeans and fewer for, say, lettuce. Other factors are of significance as well. Soybeans are highly photoperiod sensitive, meaning multiple varieties are needed to cover the numerous maturity zones that cover US soybean production areas.

Impacts on Private Investment

As the principal justification for patents, PVP and similar instruments are the stimulation of investments in research, it is important to examine the record to determine the actual effects. Theoretical analysis are generally supportive of the inducement of additional investment in research and development (R&D), but the details of a system can have significant effects on the outcome. For example, patents that are very broad can stifle subsequent research, while narrow patents may encourage additional investment, but only in an effort to 'invent around' existing patents. Patents may also be used as 'tying' goods to extend the patent-based monopoly to unrelated products. In a famous (if unrelated to plants) example, a company which owned patents on canning equipment required customers to purchase the tinned metal from the same company, which, among other things, stifled investment in improving the metal formation process by limiting market access. Many of these matters, though, also relate to the social benefits from IPR systems; here we are concerned with the simpler issue of effects on R&D investments.

Linking IPRs and R&D none the less presents some substantial economic analysis problems. This is particularly true for patents, which have existed for more than two centuries. Attributing the effects of the patent system then requires determining what would have transpired under a very different economic system. For that reason, one of the stronger pieces of evidence available relates to the effect of PVP systems which, by providing recent (1970 in the USA) protection for a group of products (self-pollinated plants) not previously protected, allows for a 'before and after' form of evaluation. That evidence is examined below. For patents, repeated surveys of firms have indicated that the existence of patents is rated low overall as a stimulus to R&D investments, below the use of trade secrets and rapid innovation, for example. There are exceptions, notably for living organisms and pharmaceuticals. Pharmaceuticals have the characteristic of being costly to develop and relatively easy to copy, so that legal (patent) protection is considered essential. Deolalikar and Evenson (1990, p. 237), for example, found pharmaceutical R&D in India fell by 40% following the patent law revision reducing the term of protection to 7 years. Canada for its part extended protection to pharmaceuticals only in 1985, it has been reported, as part of an agreement to attract R&D to Canada. Pharmaceuticals, though, have the attribute of a long and costly review process said to absorb the majority of the estimated $500 million presently required to bring a successful product to market. Pharmaceutical patents are also narrow in the sense of being applied a single compound, the one to undergo the safety and efficacy trials (Grubb, 1999). Hence, in the case of pharmaceuticals, the regulatory process is a major component of the reliance on patents.

The need for IPR protection for living organisms is based on the attribute of the organisms to reproduce. Copiers of those products need no specialized skills, and indeed farmers reproduce the product with the mere act of growing the crop. Hence, the R&D inducement effect of IPR protection for living organisms should be particularly strong. We now turn to an examination of the evidence.

Prior to the passage of the US PVPA in 1970, the breeding of open-pollinated varieties (soybeans, wheat, cotton, etc.) was completely dominated by the public sector, particularly the Land Grant Colleges. Butler and Marion (1985, Table A15) found that the investment in private sector R&D rose most rapidly during the 1966–1970 period, presumably in anticipation of the passage of the PVPA. For firms with breeding programmes in place prior to 1960 (and hence largely involved in producing hybrid maize), deflated breeding investment for 14 firms rose from $1 million in 1960 to $8 million by 1970. The 44 firms that started breeding programmes after 1960 invested $4 million by 1965 and over $8 million by 1970. Other measures of investments like numbers of breeders rose as well. Perrin *et al.* (1983, Table 3), using a separate survey, broke out investments by crop. For soybeans, only one private firm had a soybean breeder (and a $2000 budget) in 1960, a figure that rose to eight firms and $270,000 by the passage of the PVPA in 1970 and up to $4.3 million by the end of the 1970s. Those expenditure increases are substantial, even when corrected for inflation over the period. Soybeans, due to their high crop value and photosensitivity, as noted above, were an obvious market for the private sector. Investments in vegetable breeding also rose sharply, particularly following 1980, when more products were protectable under the PVPA. Other cases are not as attributable to legal protection. Investment in hybrid maize breeding rose most sharply over the 1960–1980 period, even though hybrids were specifically excluded from protection under the PVPA until 1994 and anyway have a form of 'biological' protection which is greater than the legal protection provided by PVP. Presently, most soybean varieties are produced by the private sector. In that and other crops with a major private sector presence, the public sector has focused on pre-commercial or pre-development breeding. Yet, the public sector still dominates wheat breeding and that of several minor agronomic crops like barley.

Few studies have attempted to document whether privately bred varieties have superior yields, beyond the simple assumption that they sell, so must be productive. Alston and Venner (1998) did contrast the performance of public and private wheat varieties using both state average and experimental test results. They found no significant differences, and concluded the PVPA has served primarily as a 'marketing tool'. In general, one could say there is no reason why the public sector could not continue to produce meritorious varieties, as in the pre-PVPA days for most crops, but that the legislation does stimulate private investment when public budgets decline.

More from a public welfare than investment perspective, several studies attempted to determine if the existence of PVP led to higher seed prices. Based on a survey of public sector breeders, Butler and Marion (1985, pp. 60–61) concluded the PVPA has led to 'slight to substantial increases in prices of protected varieties'. The pertinent issue, though, is if the increase is justified by higher returns, due to improved yields and/lower costs. There, it was found the benefits of improved varieties are shared between farmers and seed companies, so that farmers on average pay only 25–50 cents per dollar of expected return for improved seed.

In general, seed companies compete on market share more than price. Rising market shares of course mean greater concentration, which, with the number of mergers of seed companies, has led to very high levels of concentration in some crops.

Whether the mergers themselves are associated with IPRs is a matter of ongoing debate since the mid-1980s. IPRs created the potential for greater profits and interest in merger activity. That motivation seems to have spurred the first round of seed industry mergers in the 1980s (see Butler and Marion, 1985). A second round followed the emergence of agbiotech in the 1990s. James (2000, Table 8) lists 12 major mergers occurring in 1999 alone. Monsanto presently has an estimated 90% market share for herbicide resistance (RoundupReady) and *Bt* (YieldGard) technologies, which have been licensed to most seed companies as well as being available from Monsanto's subsidiary DeKalb. Yet the causality of the merger interest is debatable. Alternative motivations which have been suggested include: (i) maintaining access to delivery channels (seeds) for agbiotech products, and (ii) capturing stock appreciation value of seed company stock created by the advent of biotechnology applications. The latter was facilitated by the use of appreciated stock prices in stock-financed mergers.

Concentration activity though is not completely irreversible. The year 2000 saw several divestitures of agbiotech firms, such as Monsanto being partially spun off from Pharmacia, Aventis divested its agricultural holdings, and Syngenta was divested from Novartis and AstraZeneca (James, 2001). Those structural changes separated the (more profitable) pharmaceutical operations from the (less profitable) agricultural ones. The overall result, though, is a significantly more concentrated seeds industry than when public breeding dominated the sector.

Evidence on the effects of PVP in other countries, particularly developing countries, is more limited, but what is available is generally consistent with the US experience. Van Der Walt (1995, Table 5) reports that the number of registered soybean varieties jumped from six in 1975 to 23 in 1980 and 48 in 1994. PVP legislation was adopted in 1976. The most complete accounting though comes from a 1994 study of three Latin American countries (Argentina, Chile and Uruguay) with PVP protection in place (Jaffé and van Wijk, 1995). Chile is said to have adopted the legislation in order to have quick access to fruit and vegetable markets popular in North America, which Chile served on a counter-seasonal basis. At that time, US law did not prevent the importation of products of varieties used without permission, but the Chilean industry had determined that indirect acquisition of the new varieties, while less costly in a cash flow sense, led to access delays which were more costly overall.

Argentina provided the clearest evidence of the incentive effect of PVP. There, private investment held constant in a recessionary period, but that occurred only after the court processes were changed, making it possible to enforce PVPs. Clearly, IPRs are not honoured unless they are enforceable, as the Argentine case highlights.

Impacts on Access

The secondary role of IPR is the enhancement of access to products and technologies protected elsewhere. Conceptually, firms will be less willing to transfer to countries lacking effective protection so that transfer will be hindered or at least delayed. The Chilean example (see above) is one such documented case; Canada was also said to have adopted PVPs (in 1990) to facilitate access to potato varieties (Young, 1989). However, with seeds, like many products, preventing transfer is not always possible, so that the owner may choose to sell in insecure markets on the expectation that some revenue is preferable to none. That is the position frequently taken by pharmaceutical companies as well (see survey results in Lesser, 2001).

One source of evidence on the overall effect of IPR on access is through studies of the 'strength' of a national IPR system on foreign direct investment (FDI) and imports. The two measures reflect alternative means of product access, although of course much FDI and imports are independent of IPR matters. In order to conduct the analysis, it is necessary to construct an IPR scoring system. Two approaches have been used, one based on the content of national legislation and the second on interviews with practitioners.

Empirical studies generally support an expectation that stronger IPR protection does indeed enhance both FDI and imports. Results of several studies across time and models do, however, vary according to (literature review in Maskus, 2000, Chapter 4):

- production for export or domestic consumption;
- degree of industrialization – more industrialized nations show a more pronounced response to the strength of IPR;
- sector – more technologically advanced sectors (like electronics) are more IPR dependent than less technical ones (like food processing);
- exchange rate; and
- stock of existing FDI – countries with significant prior FDI tend to attract more.

Unresolved is the issue of whether the IPR index measures the role of IPRs *per se*, or whether it serves as a proxy for the existence of a number of factors like effective courts that are conducive to investments. There has not been, nor is there likely to be, a final resolution to that issue. However, studies that examine countries of differing technological sophistication show differential responses to measures of IPR strength, independent of other measures of institutional efficiency. This suggests that the IPR system is the cause of FDI, and not vice versa.

Effects of IPR on Research

Eisenberg (1989) considered a number of theories of innovation and scenarios of science and relationship between researchers and rights owners. The results of her analysis are not unambiguously in support of free research access. 'But as the line between basic and applied research becomes blurred in certain fields, patent protection increasingly threatens to encroach on the domain of research science, making it necessary to work out an accommodation ...' (Eisenberg, 1989, p. 1986). Biotechnology is one of those areas where applied/basic research boundaries are blurred. Heller and Eisenberg (1998) write of biomedical research where both a 'proliferation of patents on individual fragments held by different owners' and 'stacked' licence fees 'may lead paradoxically to fewer useful products'. In agbiotech, a frequently cited example is Golden Rice where between 0 and 44 separately protected components have been identified, the actual number depending on which country is being considered (Kryder *et al.*, 2000).

Some distinction should be made between these concerns and the research exemption *per se*. The research exemption takes on different forms for different protection mechanisms. With PVP it is statutory – written into the legislation – while in patent law there is typically no text so the interpretation is based on case law. Research access itself is, however, rarely of concern – what is problematic is securing rights *ex post* to commercializing products of that research as a classic 'hold-up' case ensues. A hold-up applies to a case where power is used to extract a better agreement when the other party has invested considerable resources and has few options but to acquiesce. The negotiation of commercialization terms *ex ante* would reduce the hold up component, but has the practical problem of imposing large transaction costs for many projects that will never materialize. Negotiations are seen as particularly burdensome for public sector researchers due to smaller staff resources with the needed expertise and a less commercial outlook overall. Hence, while a statutory research exemption in patent law would help to clarify rights and facilitate access in some cases, the issue of securing rights to commercialize products developed under a research exemption is a distinct and separate one.

Conclusions: Projecting Effect on to Animal Patents

Our knowledge of the ongoing effects of IPR protection for plants remains fragmentary. Projecting from that base to likely future effects of IPRs on the large and diverse animal sector is that much more uncertain. Yet many of the applications to plants can be understood (at least after the fact) as direct consequences of economic behaviour on the part of researchers and users. Hence, at least for the animal groups that are used in a commercial environment, it is possible to draw on experiences with plants to project some likely future adjustments with animals.

Prices

In general, patents for animals can be expected to raise prices and increase sector concentration. In practice, the magnitude of the effects will be sector-specific. Some, from developers to broiler chicken genetics producers, were highly concentrated without patents. Others, like purebred cattle breeders, are highly dispersed and could become concentrated with active use of patents. Price effects are affected by: (i) competition within the sector, and (ii) the sector profitability. For example, the high-margin pharma sector can sustain greater price increases than low-margin livestock operations. Hence, no general projection is possible, but the situation should be monitored carefully by participants and regulators alike.

Investment in R&D

The application of IPRs does increase R&D investments for living organisms, but the magnitude is dependent on several factors. Large impacts are likely only if the market for the product is substantial. The product must demonstrate a clear positive response to the investment, and it must be easily copyable and not readily protectable by trade secrets (secrecy). Thus, animal disease models for minor illnesses with small research efforts would be expected to respond little to patents. High-valued uses, such as the production of pharmaceuticals in mammary glands, which can be managed in a confinement/contract production environment, would also show limited response. Synthetic breeds that have biological protection (including about half the feeder hogs sold and essentially all poultry stock) would also respond little. Finally, the costs of enforcement and royalty collection must be considered so that complex systems such as will be needed for the dispersed cattle sector will return relatively less to investors, reducing the incentive. These factors explain in part why animal patenting has been focused on relatively few classes and use areas compared to the diversity of commercial uses of animals.

Effects on technology transfer

Experiences with plants shows that IPR protection in general enhances the availability of products in countries where protection is allowed and received. That is, countries where plant patents or PVP are held typically can access products more readily than when IPR protection is not available. Note a distinction must be made between the initial accessibility of an invention and the speed of its dispersion. IPRs generally increase the speed of initial access but delays dispersion due to the higher price associated with IPRs. Some caveats must be added to these general statements, including the degree to which the owner can prevent transfer and alternative protection mechanisms.

For animals, the same general conditions would apply. The movement of (say) companion dogs across national borders would be difficult to control, so that a patented companion dog might be transferred even to countries where patent protection is not held, and indeed may not be available. Synthetic chicken breeds, in contrast, while easily moved, are not readily copied so they too are generally available (but typically as production rather than parent or grandparent stock). Disease models are relatively controllable given the specialized nature of the market, but once available it would be difficult to prevent further dissemination.

The preceding points are particularly relevant for animals due to the ambiguity/limited extent of nations that provide patent protection. Financial incentives for researchers are the product of price times market size times market duration (length of time until the product becomes outdated or new competition emerges). Holding price constant, there is a direct trade-off between market size and duration; short duration markets can be offset by market size. One dimension of market size is of course the number of national markets available, so to the extent that patents are not available in most nations, the incentives for investing in animal research will be limited for products lacking international markets.

Effects on research

Protected plant products contain not one but often dozens of patented components. Additional parts may have been made available under a Material Transfer Agreement (MTA) for research purposes only. When the time comes for commercializing the new composite product, the absence of freedom to operate has been recognized as a real barrier. Since patented animals share many of the multi-component aspects as plants, the same impediment to commercialization can be anticipated. Careful planning and records management, at a minimum, will minimize the problems encountered.

Plants and animals also share the characteristic of the need for access to existing genetic resources for successive improvements. However, a patent can prevent the use of a patented plant or animal for improvement breeding purposes. This can severely limit breeding advancements, to the detriment of competitors and the general public. The practical effects are, however, limited if the parent stock remains in the public domain; breeders can return to the parents for subsequent breeding work. Yet that approach reduces the cumulative aspects of breeding due to the necessity to return to a previous generation for selecting sires. Of less concern is the patenting of a line with a particular trait. In those cases, the patent will typically read to 'a ... with the characteristic of ...'. Effectively, what is protected is the trait, not the vast amount of germplasm untouched by the researcher. However, if the patent is for the next iteration, as is now being done with maize and soybeans (see above), then the researcher can withhold all the germplasm in the patented animal from subsequent breeding – the case identified above. Because there is, in the USA as elsewhere, no alternative mechanism but patents for protecting animals, the extension of patents to animal lines with limited inventive step, following what has already occurred with plants, seems highly likely. The development of 'animal breeders' rights' legislation, granting a research exemption, for protecting iterative improvements in livestock breeding would resolve the research access issue (see Chapter 1). For more fundamental development, including transgenic organisms, patents seem well suited.

References

Alston, J.M. and Venner, R.J. (1998) The effects of U.S. Plant Variety Protection Act on wheat germ plasm improvement. Paper presented at the symposium, IPR and Agricultural research Impact, CIMMYT, Mexico, 5–7 March 1998.

Butler, L.J. and Marion, B.M. (1985) *The Impacts of PATENT Protection on the U.S. Seed Industry and Public Plant Breeding*. University of Wisconsin, N.C. Project 117, Monograph 16, September.

Deolalikar, A.B. and Evenson, R.E. (1990) Private inventive activity in Indian manufacturing: its extent and determinants. In: Evenson, R.E. and Raines G. (eds) *Science and Technology: Lessons for Development Policy*, Chapter 10. Westview Press, Boulder, Colorado.

Eisenberg, R.S. (1989) Patents and the progress of science: exclusive rights and experimental use. *University of Chicago Law Review* 56, 1017–1086.

Foster, W.E. and Perrin, R. (1991) Economic Incentives and Plant Breeding Research. Department of Agriculture and Resource Economics, North Carolina State University, DARE: 91–05, May.

Grubb, P.W. (1999) *Patents for Chemicals, Pharmaceuticals and Biotechnology*. Clarendon Press, Oxford.

Heller, M.A. and Eisenberg, R.S. (1998) Can patents deter innovation? The anticommons in biomedical research'. *Science* 280, 698–701.

Jaffé, W. and van Wijk, J. (1995) The impact of plant breeders' rights in developing countries. University of Amsterdam, Inter-American Institute For Cooperation on Agriculture, October.

James, C. (2000) Global Status of Commercialized Transgenic Crops: 1999. ISAAA briefs No. 17–2000. ISAAA, Ithaca, New York.

James, C. (2001) Global Review of Commercialized Transgenic Crops: 2000. ISAAA briefs No. 23–2001. ISAAA, Ithaca, New York.

Kryder, R.D., Kowalski, S.P. and Krattiger, A.F. (2000) *The Intellectual and Technical Property Components of pro-Vitamin A Rice (GoldenRice). A Preliminary Freedom-to-operate Review.* ISAAA Briefs No. 20, Ithaca, New York.

Lesser, W. (1987) The impacts of seed patents. *North Central Journal of Agricultural Economics* 9, 37–48.

Lesser, W. (1994) Valuation of Plant Variety Protection certificates. *Review of Agricultural Economics* 16, 231–238.

Lesser, W. (2001) Role of IPR in biotechnology transfer – corporate views. Available from www.WIPO.org under studies on intellectual property.

Maskus, K.E. (2000) *Intellectual Property Rights in the Global Economy.* Institute for International Economics, Washington, DC.

Perrin, R.K., Hunnings K.A. and Ihnen, L.A. (1983) *Some Effects of the U.S. Plant Variety Protection Act of 1970.* Department of Economics and Business, North Carolina State University, Economic Research Report No. 43, August.

Seay, N.J. (1993) Intellectual property rights in plants. In: Baenziger, P.S., Kleese R.A. and Barnes, R.F. (eds) *Intellectual Property Rights: Protection of Plant Materials,* Chapter 6. CSSA Special Publication No. 21, Crop Science Society of America, Madison, Wisconsin.

Van Der Walt, W.J. (1995) *Brief Review of Intellectual Property Rights in South Africa.* Mimeo, March.

Young, S. (1989) Testimony on Bill C-15, An Act Respecting Plant Breeders Rights. Issue No. 5, House of Commons, Nov. 2.

8

Patenting and Sequencing the Genome

Heidi S. Nebel, Lila A.T. Akrad and Tonya Y. Brady

McKee, Voorhees & Sease, 801 Grand Avenue, Suite 3200, Des Moines, IA 50309-2721, USA

Today, the only wealth there is in the world is the wealth that comes from the human mind. (Olsen, 1997, quoting Marsa, 1995, p. 36)

Introduction

The field of biotechnology has progressed extensively with the completion of the human genome project, the success of animal cloning and the development of human embryonic stem cells. 'While the academic and industrial research institutions in the United States are world-wide leaders in the development of biotechnology, inventions generated by these entities must be adequately protected to ensure continued innovation' (Sampson, 2000, p. 1234). The worldwide desire to promote innovation, however, must be balanced with the challenges that biotechnology poses to the US patent law system. The challenge for the US Patent & Trademark Office (USPTO) and the Federal Circuit has been to promote innovation while not granting overly broad patent rights to the inventors.

The US Constitution provides that Congress shall have the power '[t]o promote the Progress of Science and the useful Arts, by securing for limited Times to Authors and Inventors the exclusive Right to their respective Writings and Discoveries' (US Constitution). However, when it comes to living organisms, this power is strongly debated.

The biotechnological revolution of the 1980s and 1990s has enabled scientists to isolate genetic material of living organisms and to induce precise modifications so that organisms portray and carry desired genetic traits (Scalise and Nugent, 1995). However, it was not until the paramount US Supreme Court 5–4 decision in 1980, *Diamond vs.*

Chakrabarty (1980), that the Court interpreted patent laws to include genetically altered living organisms as patentable subject matter. In 1988, the USPTO granted the first patent on a 'non-naturally occurring non-human multicellular living organism': a mouse genetically engineered to be particularly susceptible to cancer (Patent & Trademark Office Notice, 1987; US Patent No. 4,736,866, 1988). The USPTO's decision to grant the mouse patent cleared the way for scientists and industry to file patent applications for bioengineered animals and plants, provided that they met the traditional standards set forth in 35 U.S.C. §§101, 102, 103 and 112.

One important development through the evolution of biotechnology and patent law is that the issue is no longer whether animals and plants can be patented, but which animals and plants should receive patent protection (Iwasaka, 2000). Today, what is being patented is the process of discovering and isolating in the laboratory certain strings of DNA that were not obvious before. The patent covers only the process, not the *in situ* raw material.

This chapter focuses on genomics and patenting the results of this field as well as the effects on the biotechnology industry in acquiring gene patents in the future. After a general examination into the US patent law, the next section will include a brief overview of biotechnology and some scientific fundamentals. Then the Human Genome Project will be discussed and how the recent discoveries may be influenced with respect to the US patent laws. This will then be followed by the next section, which will briefly explore some of the ethical issues surrounding patent law. Finally, this chapter will conclude with a synopsis of the legality of the USPTO guidelines and the possible affect it will have on the future of biotechnological innovations (Hoffert, 1998).

USPTO

The foundation of the US patent law is rooted in Article 1, Section 8, Clause 8 of the US Constitution. The Constitution broadly empowers Congress '[t]o promote the Progress of Science and useful Arts, by securing for limited Times to Authors and Inventors the exclusive Right to their respective Writings and Discoveries' (US Constitution). This progress is promoted through the enactment of patent laws, which offer inventors the exclusive rights for a limited period of time as an incentive for their ingenuity and research efforts (*Diamond vs. Chakrabarty*, 1980). Thomas Jefferson authored the first of these US patent laws, the Patent Act of 1793 (Act of 21 February 1793). The current Patent Act is codified under Title 35 of the US Code and grants the inventor a 20-years exclusive right to the invention (35 U.S.C.A. §154 (a)(2), 2001). Inherent in its basic mode of award, a brief monopoly, is the concept that inventors will be given the time to exploit their creation (*Report of the President's Commission on the Patent System*, 1966). It is not necessary for them to fear the more powerful or financially sound competitors, for their power is to exclude all others from making, using, or selling the patented concept (*Report of the President's Commission on the Patent System*, 1966). In exchange for this grant, the inventor must disclose the invention in what is designated as the 'enabling disclosure' (35 U.S.C.A. §112, 1984).

The issuance of a patent is dependent on fulfilling the requisite statutory requirements. The threshold statutory inquiry involves whether the claimed invention contains patentable subject matter (35 U.S.C.A. §101, 1984). Although the claimed invention may contain patentable subject matter, the invention must also satisfy a three-pronged inquiry of novelty, non-obviousness and utility before a patent will be issued (35 U.S.C.A. §101 at Part III, 1984). While all of these patentability requirements have evolved unique characteristics in the area of biotechnology, perhaps the most dramatic development has occurred under 35 U.S.C. §112, first paragraph.

Until 1930, patents on plants and animals were barred (*O'Reilly vs. Morse*, 1853). This prohibition was based upon three main assumptions. First, patents on animals and plants would allow humans to hold exclusive rights to other living organisms (Singer, 1990). Second, some argued that granting such patents would unjustly enrich those who were simply the first to apply for a patent on a naturally occurring organism (*O'Reilly vs. Morse*, 1853). Third, some invoked a Lockean philosophy of labour and just desserts in contending that live-organism patents were inappropriate because they did not require the patent holders to mix their labour with the naturally occurring item (Brody, 1991).

Today, patentability of claims is supported if they meet the criteria of 35 U.S.C. §§101, 102, 103, without regard to whether the claimed subject matter is alive (*Application of Bergy*, 1979). To begin to understand where intellectual property protection of biotechnology is today, we must look back at how far it has come. In 1972, Chakrabarty, a microbiologist, filed a patent application based upon his discovery that a bacterium of the genus *Pseudomonas*, when genetically engineered, was capable of successfully 'eating' oil spills (*Diamond vs. Chakrabarty*, 1979). This property was possessed by no known naturally occurring bacteria. As part of his invention, Chakrabarty claimed the genetically altered bacterium. The claims were initially rejected by the USPTO. Chakrabarty appealed, and the Patent Office Board of Appeals affirmed (*Diamond vs. Chakrabarty*, 1979). The Court of Customs and Patent Appeal (CCPA) reversed on the authority of its prior decision of *In re Bergy*, which held that whether microorganisms are alive is without legal significance for purposes of patent law (1977). After the Supreme Court vacated

Bergy, the CCPA reaffirmed the patentability of the microorganism claims in *Chakrabarty* (1977). The Supreme Court granted *certiorari* in both cases (*Diamond vs. Chakrabarty*, 1979). At the time of the decision, only *Chakrabarty* was left. The Court stated: '[T]he patentee has produced a new bacterium with markedly different characteristics from any found in nature, and one having the potential for significant utility. His discovery is not nature's handiwork, but his own; accordingly, it is patentable subject matter under §101' (*Diamond vs. Chakrabarty*, 1979). The Supreme Court summarily rejected the argument that the 1930 Plant Patent Act was the sole means of protection of discoveries concerning living organisms (*Diamond vs. Chakrabarty*, 1979). Social and public policy concerns militating against the patentability of living organisms were similarly rejected:

> The legislative process, the [government] argues, is best equipped to weigh the competing economic, social and scientific considerations involved, and to determine whether living organisms produced by genetic engineering should receive patent protection. In support of this position the [government] relies on our recent holding in *Parker vs. Flook* … and the statement that the judiciary must proceed cautiously when … asked to extend patent rights into areas wholly unforeseen by Congress (*Diamond vs. Chakrabarty*, 1979, p. 314).

> It is, of course, correct that Congress, not the courts, must define the limits of patentability; but it is equally true that once Congress has spoken it is 'the province and duty of the Judicial Department to say what the law is.' Congress has performed its constitutional role in defining patentable subject matter in §101; we perform ours in construing the language Congress has employed. In so doing, our obligation is to take statutes as we find them, guided, if ambiguity appears, by the legislative history and the statutory purpose. Here, we perceive no ambiguity. The subject-matter provisions of the patent law have been cast in broad terms to fulfill the constitutional and statutory goal of promoting 'the Progress of Science and the useful Arts' with all that means for the social and economic benefits envisioned by Jefferson. Broad general language is not necessarily ambiguous when congressional objectives require broad terms (*Diamond vs. Chakrabarty*, 1979, p. 315).

> It is argued that this Court should weigh [the] potential hazards in considering whether respondent's invention is patentable subject matter under §101. We disagree. The grant or denial of patents on microorganisms is not likely to put an end to genetic research or to its attendant risks. The large amount of research that is already occurred when no researcher had sure knowledge that patent protection would be available suggests that legislative or judicial fiat as to patentability will not deter the scientific mind from probing into the unknown any more than Canute could command tides. Whether respondent's claims are patentable may determine whether research efforts are accelerated by the hope of reward or slowed by want of incentives, but that is all (*Diamond vs. Chakrabarty*, 1979, p. 317).

In short, with elegance, style and the stroke of a pen, Chief Justice Berger's opinion in *Diamond vs. Chakrabarty* brought the patent law to a new level. The federal patent laws were about to be extended into unchartered territory. Presently, the patent system has begun to be fully applicable to the emerging possibilities of biotechnology. The question remains as to the outcome of such a path.

It was in *Brenner vs. Manson* that the US Supreme Court set forth criteria for patent utility (1966). The Court in concluding that the invention failed to meet the utility requirement promulgated three essential findings affecting biotechnology applications today (*Brenner vs. Manson*, 1966). First, the Court determined that Congress did not intend the term 'useful' to be so broad as to include 'any invention not positively harmful to society' (*Brenner vs. Manson*, 1966, p. 533). Second, the Court expressed concern regarding the issuing of a monopoly for a compound with an unknown function (*Brenner vs. Manson*, 1966). Finally, the Court concluded that utility is not established by merely proving that the product is the result of scientific research (*Brenner vs. Manson*, 1966).

The utility requirement does present problems in the patenting of gene sequences (Michaels, 1994). In particular, the attempted patenting of expressed sequence tags (ESTs), by the National Institutes of Health (NIH), discussed below, is illustrative of the controversy of the utility requirement in the biotechnological arts (Burchfiel, 1995). Due to complaints of the USPTO rejection of biotech patents due to lack of utility, the USPTO issued Utility Guidelines (USPTO, 2000). The 2001 Utility Guidelines require that a claimed invention either have a 'well-established' utility or assert a 'specific, substantial, and credible' utility (Utility Examination Guidelines, 2001, p. 1098). The guidelines state that the requirement of specific and substantial utility 'excludes throw-away, insubstantial, or nonspecific utilities, such as the use of a complex invention as landfill' (Utility Examination Guidelines, 2001, p. 1098). These guidelines help to establish procedures for the examiners to follow in their review of the patent applications and specifically set forth the criteria to determine 'the evidence sufficient to establish a lack of utility' (USPTO, 2000). Congress however, has continued to authorize the USPTO with the unenviable job of anticipating how the courts will decide cases involving new technologies.

The utility standard under 35 U.S.C. §101 has long been considered a de minims one and is seldom a real bar to the issue of a given patent within the realm of biotechnology (*Ex parte Drulard*, 1983). While no court has yet ruled directly on the utility of DNA compositions, the Board has indicated that the practical utility standards apply to biotechnology patents. In *Ex parte Maizel*, the Board held that the lack of an asserted use for a claimed growth factor protein, while not an issue in controversy, may have indicated that the protein lacked practical utility (1992). In *Ex parte Deuel*, the Board questioned whether or not a DNA sequence derived from rat was sufficiently homologous to human DNA to be used as a probe for similar DNA sequences (1993).

Section 102 requires novelty of the claimed invention (35 U.S.C.A. §102, 1984). As delineated in §102 of the Patent Act, an invention that does not exist in the prior art is novel (35 U.S.C.A. §102(a), 1984). In other words, the Patent Act requires an applicant to be the 'first inventor to confer the benefit of the invention on the public' (Kight, 1998, p. 1008). This requirement mandates that the *quid pro quo* for the limited monopoly of a patent is the development or revelation of something that is truly new. Today some argue that it is incorrect to allow any living matter to be considered novel because gene sequences exist naturally (*Funk Brothers Seed Co. vs. Kalo Innoculant Co.*, 1948). Yet with the progression of biotechnology today, organisms are altered in such a way that they are dramatically different from 'naturally occurring' organisms. As science has advanced, so has the law. Thus, genes in an isolated or purified form may be patented, but a gene sequence that remains in the form in which the scientists initially discovered the sequence is not patentable under current law (Erramouspe, 1996).

The next hurdle encompasses the non-obviousness requirement of the Patent Act under §103 (35 U.S.C.A. §103, 2001). The analysis under the non-obviousness issue involves the examination of whether 'the subject matter as a whole would have been obvious at the time the invention was made to a person having ordinary skill in the art to which said subject matter pertains' (35 U.S.C.A. §103(a), 2001). In the biotechnology field, obviousness has been a difficult subject, because researchers in isolating different gene sequences use similar techniques, although the gene sequence is itself new (Eisenberg, 1990).

The Federal Circuit decision of *In re Deuel* specifically addressed the non-obviousness requirement in the context of biotechnology (1995). The court appeared to loosen the modern test for obviousness determined by the Supreme Court in *Graham vs. John Deere Co.* (1966), by asserting that '[a] general motivation to search for some gene that exists does not necessarily make obvious a specifically-defined gene that is subsequently obtained as a result of that search' (*In re Deuel*, 1995, p. 1558). The decision of *In re Deuel* allowed for DNA molecules to be patented even if the inventor discovered the DNA by using an 'obvious' scientific method (1995). None the less, in 1995 an amendment to §103 suggests that both the process and the subject matter must be non-obvious in order to meet the non-obviousness requirement (35 U.S.C.A. §103(b)(1), 2001).

Today, *Graham vs. John Deere Co.* remains the basis for the modern test of obviousness (1966). The US Supreme Court articulated the three requirements: (i) the courts must determine the scope and content of the prior art; (ii) the courts must ascertain the differences between the prior art and the claimed invention; and (iii) the courts must resolve the level of ordinary skill in the relevant art (*Graham vs. John Deere Co.*, 1966). Furthermore, the courts must consider obviousness or non-obviousness of the subject matter in light of all of these elements (*Graham vs. John Deere Co.*, 1966). The Federal Circuit later added a fourth element to the three-pronged *Graham* test, requiring consideration of the product's success in the commercial market, the existing demand for the product and the inability of others to create the invention (*Hybritech, Inc. vs. Monoclonal Antibodies, Inc.*, 1986). Thus, by applying the above test a claimed invention is rendered obvious if there is a nexus between the invention and the prior art that could be accomplished by an individual of ordinary skill in the art (35 U.S.C.A. §103(a), 2001).

In 35 U.S.C. §112, the first paragraph sets forth the statutory requirements for the 'written description,' 'enablement' and 'best mode' requirements for patentability (35 U.S.C. §112(1), 1984). The written description requirement has been the subject of much debate. The story of the modern written description requirement began in *In re Ruschig*. The Court of Customs and Patent Appeals, the Federal Circuit's predecessor, announced in *Ruschig* a separate written description requirement designed to ensure that the applicant was in possession of the claimed invention at the time the patent application was filed (*In re Ruschig*, 1967). The court later clarified that while the enablement requirement teaches how to make and use an invention without undue experimentation, the written description requirement serves 'to put the public in possession of what the party claims' as its invention (*Vas-Cath, Inc. vs. Mahurkar*, 1991). Although patent law generally allows an inventor to patent an invention that has not yet been reduced to practice, the Federal Circuit has essentially disallowed this practice in the 'unpredictable art' of biotechnology by using a heightened written description requirement (Mueller, 1998). Therefore, an inventor must first make or use the invention before a patent application is filed on a biotechnological invention (*Burroughs Wellcome Co. vs. Barr Laboratories, Inc.*, 1994).

Another pertinent case decided by the Federal Circuit is *Fiers vs. Revel* that also affected the written description requirement (1993). The *Fiers* decision involved a three-way interference proceeding among three foreign inventors involving DNA coding for human fibroblast beta-interferon polypeptide (β-IF) (1993). On appeal, Fiers asserted that the stringent written description requirement set forth in *Amgen, Inc. vs. Chugai Pharmaceutical Co.* only applies when the disclosed method for isolating a DNA sequence could not be easily carried out by one of ordinary skill in the art (*Fiers vs. Revel*, 1993). Further, Fiers argued that *Amgen* allows a conception of a DNA sequence by its method of isolation (*Fiers vs. Revel*, 1993). The Federal Circuit rejected both of these arguments noting that Fiers inappropriately focused on the issue of enablement rather than written description (*Fiers vs. Revel*, 1993). The court stated that 'adequate written description of DNA requires more than a mere statement that it is part of the invention and reference to a potential method for isolating it; what is required is a description of the DNA itself' (*Fiers vs. Revel*, 1993). Thus, after *Fiers*, an inventor must disclose a specific characteristic of the claimed DNA sequence sufficient to convey to one skilled in the art that the inventor was in possession of the invention at the time the patent application was filed (1993).

The Federal Circuit most recently addressed the issue of the written description requirement for DNA inventions in *Regents of the University of California vs. Eli Lilly & Co.* (1997). This decision has created not only a great deal of controversy but also uncertainty with regard to the scope and validity of biotechnology patents (Stewart, 1999). The Federal Circuit rejected UC's argument that the disclosure contained sufficient written description of the human insulin cDNA because the examples in the disclosure described how to isolate the cDNA (*Regents of the University of California vs. Eli Lilly & Co.*, 1997). The court simply stated 'enabling a person skilled in the art to obtain a DNA sequence does not sufficiently describe that sequence' (*Regents of the University of California vs. Eli Lilly & Co.*, 1997, p. 1567). The court further explained that unless an inventor possesses the complete sequence of a gene or cDNA, that inventor cannot visualize or recognize its identity (*Regents of the University of California vs. Eli Lilly & Co.*, 1997). Therefore, even if the disclosure is sufficient to render the invention non-obvious, it still may be insufficient to satisfy the written description requirement (*Regents of the University of California vs. Eli Lilly & Co.*, 1997).

Furthermore, the enablement requirement's crucial language in 35 U.S.C. §112 mandates that the specification of a patent teaches a person skilled in the art how to make and use the full scope of the invention without 'undue experimentation' (*Genetech Inc. vs. Novo Nordisk A/S*, 1997). Thus, the claims of the patent application must be enabled by the specification at the time the application was first filed (*Hybritech Inc. vs. Monoclonal Antibodies, Inc.*, 1986).

In *Amgen, Inc. vs. Chugai Pharmaceutical, Co.* the evolution of 35 U.S.C. §112, first paragraph, the enablement requirement, began to develop (1991). The patents at issue in *Amgen* involved technology relating to a way to purify and isolate DNA sequences encoding for production of human erythropoietin (EPO), a protein used to therapeutically stimulate the production of red blood cells for treatment of anaemia and other blood disorders (1991). Amgen sued Chugai for patent infringement, and the defendants counterclaimed that Amgen's patent was invalid (*Amgen, Inc. vs. Chugai Pharmaceutical, Co.*, 1991). The Federal Circuit affirmed the district court's finding that Amgen's patent was invalid for lack of enablement (*Amgen, Inc. vs. Chugai Pharmaceutical, Co.*, 1991). The problem for Amgen was their broad claim to EPO analogues. This claim encompassed an astronomical number of possible DNA sequences without the ability to predict the outcome of their encoded proteins (*Amgen, Inc. vs. Chugai Pharmaceutical, Co.*, 1991). Amgen argued that its data on 50–80 EPO analogues was sufficient to show enablement of its claims (*Amgen, Inc. vs. Chugai Pharmaceutical, Co.*, 1991). However, the court found that Amgen's claims to all EPO-like analogues were not enabled because the specification only disclosed how to make 'the gene and a handful of analogs whose activity has not been clearly ascertained' rather than a large set of EPO-like activity (*Amgen, Inc. vs. Chugai Pharmaceutical, Co.*, 1991). This decision limits an inventor's ability to protect a patented gene by claiming all possible biologically active variations of the DNA sequence of the gene. Furthermore, this case also had an impact on the written description requirement (discussed earlier), whereby an inventor claiming a DNA sequence must have an adequate conception of the DNA sequence before filing a patent application (*Amgen, Inc. vs. Chugai Pharmaceutical, Co.*, 1991).

The Federal Circuit affirmed the district court's finding in *Enzo Biochem vs. Calgene* that two patents exclusively licensed to Enzo regarding genetic antisense technology were invalid due to lack of enablement (1999). Genetic antisense technology involves the expression of an 'antisense' mRNA transcript from a specially designed DNA construct. This antisense mRNA is then complementary to the mRNA naturally expressed in a cell allowing it to bind to the naturally expressed mRNA, thereby preventing the cellular machinery from translating the natural mRNA into a protein (*Enzo Biochem, Inc. vs. Calgene, Inc.*, 1999). The patents broadly claimed the use of antisense technology in 'any organism containing genetic material which is capable of being expressed', including all prokaryotic and eukaryotic organisms (US Patent No. 5,190,931, 1993). The court stated that while the specification 'set forth the basic blueprint for the manner in which the invention might be practiced in all types of cells', the specification in reality did not enable the broad claims of the patents (*Enzo Biochem, Inc. vs. Calgene, Inc.*, 1999). The court concluded stating that Enzo's patents were a mere 'germ of the idea' and would have required undue experimentation to develop into Calgene's invention of the unique tomato (FLAVR SAVR) (*Enzo Biochem, Inc. vs. Calgene, Inc.*, 1999).

None the less, it is important for inventors and their attorneys to remember that because the chemical and biotechnological arts are considered unpredictable since scientists are not yet able to predict how simple chemical changes will affect chemical reactions or physiological reactions, the scope of enablement in the chemical and biotechnological arts varies inversely with the level of unpredictability in the art (*In re Fisher*, 1970).

Biotechnology

Biotechnology is a broad term that refers to the biological processes and recently developed technologies that actually harness the abilities of organisms, or their component parts, for the benefit of science, commerce and/or individuals (US Department of Commerce, 1992). Specifically, the principles of 'cell and tissue culture, cell fusion, molecular biology, and ... recombinant deoxyribonucleic acid (DNA) technology [are used] to generate unique organisms with new traits or organisms that have the potential to produce specific products' (*McGraw-Hill Encyclopedia of Science and Technology,* 1992). With biotechnology techniques, scientists have sought to study the biological systems of single-celled or multicellular organisms as a way to better understand more complex living systems (US Congress, Office of Technology Assessment, 1988). These techniques have today progressed into a use of microorganisms for the production of medicines, antibiotics, hormones and vitamins, among other products (US Department of Commerce, 1992).

'Biotechnology is one of the most research intensive and innovative industries in the global economy' (Feisee and Stanton, 2001). US biotechnology companies raised $33 billion dollars on Wall Street in 2000, according to Ernst & Young (LLP, 2001). Biotechnology companies use gene patents as legal protection for their innovations. Such companies contend that unless they patent genes, they cannot invest the enormous capital necessary to develop medical tests. Opponents, however, counter by arguing that the biotech industry is driven not so much on the expectation that a specific gene sequence will lead to a new drug, but more so by the idea that if enough patents are obtained on DNA sequences, one of them will be of value (Regalado, 2000).

The decision in *Amgen, Inc. vs. Chugai Pharmaceutical Co.* (1991) indicated that genes and gene sequences are patentable subject matter if they are 'novel purified and isolated sequences' derived from the original object in nature (Van Brunt, 2000). *Amgen* has the distinction of likely making genes and gene sequences patentable (Van Brunt, 2000). Gene and gene sequences are isolated and purified using a recombinant DNA cloning process. Therefore, under the *Amgen* standard, it can be argued that gene sequences are arguably sufficiently different from their natural counterparts to be patentable subject matter (Van Brunt, 2000). This conclusion is supported by the fact that patents are usually sought on cDNA sequences, which is a modified portion of the actual human DNA.

Natural compounds such as human proteins are said to occur in nature. They can be patented only if they are in some way removed from nature (*Primer: Genome and Genetic Research,* 2001). Therefore, when a protein has been purified away from the human body, a patent can be obtained on it in its purified state (*Primer: Genome and Genetic Research*, 2001). In this case, the patent does not cover the person (*Primer: Genome and Genetic Research*, 2001). In fact, the USPTO does not allow anyone to patent a human being under any circumstances (*Primer: Genome and Genetic Research*, 2001). However, isolated and purified genes are patentable inventions if they meet patentability requirements of novelty, non-obviousness and useful.

No legal distinction exists between a gene patent and patents on other chemicals (*Primer*: *Genome and Genetic Research*, 2001). There is no distinction because every gene is a molecule that can be defined as a specific structure (*Primer*: *Genome and Genetic Research*, 2001). Recall that a gene is the fundamental physical and functional unit of heredity. It is composed of polymers of DNA. DNA, in and of itself, has no functional property (*Primer*: *Genome and Genetic Research*, 2001). It is a chemical that when placed in an 'appropriate environment' will direct synthesis of specific proteins (*Primer*: *Genome and Genetic Research*, 2001). When it comes to genes, genes isolated in a laboratory setting are not considered to be in their natural form. In order to be studied, the gene is in some way altered, for example, by purification or modification (Regalado, 2000).

Gene patents have both importance and potential dangers. While gene patents provide incentive for attracting private investments in research and development of medical therapies, academic scholars and researchers alike fear that gene patents will stifle research. Many opponents to patents that include genetic information ask why would a company want to pursue researchers and block them from using diagnostic tools. A prime example of the clash between scientist and corporation came when Human Genome Sciences secured a patent on the gene that controls how AIDS begins infecting its victims (*Genetic Face-Off*, 2001). Although the biotechnology company had isolated and decoded the sequence, company officials acknowledge that at the time they applied for the patent, its use was unknown to them (*Genetic Face-Off*, 2001). However, a group of scientists claimed they were the ones who demonstrated the gene could be used to explain why some people repeatedly exposed to HIV never developed the disease (*Genetic Face-Off*, 2001). Dr Robert Gallo of the University of Maryland, who headed a group of academic researchers studying HIV infection, and other groups of independent researchers later learned that the gene was a viral receptor that the HIV virus attaches to (*Genetic Face-Off*, 2001). In addition, they also discovered that defective versions of the gene translate a protein that suppresses infection by preventing HIV from attaching to cells (*Genetic Face-Off*, 2001). The chief executive officer of Human Genome Sciences stated that the patent did not impede outside research and that the company was making the gene available to other researchers at no cost (*Genetic Face-Off*, 2001).

While the new guidelines enacted by the USPTO will require applicants to better demonstrate function and utility, others, such as officials at NIH, complain that the new guidelines fail to go far enough (*Genetic Face-Off*, 2001).

Human Genome Project

The Human Genome Project (HGP), originally designed in 1988 to be a 15-year $3 billion dollar research effort, was launched by the USA with the goal of mapping the entire human genetic blueprint – the human genome (*Hearing Charter*, 2000). Today, the HGP is an international scientific collaboration that began in 1990 to characterize the genome of humans and selected model organisms through complete mapping and sequencing of their DNA (Division of Extramural Research, 2001). This effort is believed to expand the knowledge of researchers to investigate not only genetic and inherited diseases, but to understand, on a more expansive and detailed level, specific human traits associated with particular genes at precise locations on the chromosome (Human Genome Project, 2001). In addition, the project goals are to: store this information in databases, improve tools for data analysis, transfer related technologies to the private sector and address ethical, legal and social issues (ELSI) that may arise from the project (Human Genome Project Information, 2001).

Many scientists believe that this project has the potential to revolutionize both therapeutic and preventative medicine by providing resounding insights into basic biochemical processes that underlie many human diseases (Human Genome Project Information, 2001). While the HGP may have the potential to revolutionize areas such as medicine, at the same time, it has the potential to disarrange the patent system.

The basic premise behind sequencing the human genome is that it will benefit mankind (Miller, 1997). One of the most debated issues concerning genome sequencing is the patentability of DNA. Historically, the US PTO has granted DNA patents (Miller, 1997). Many of the first genes to be patented were based on a cloning technique in which the protein was first discovered, and the gene was found by working from the protein to identifying the responsible gene (American Medical Association, 2000). This technique took years to isolate a single gene (American Medical Association, 2000). Today, there are high-speed gene sequencing techniques that allow genes as well as fragments of genes to be quickly discovered without the need of knowing the functions of the proteins produced by the gene (American Medical Association, 2000). The onset of sophisticated computer technology has allowed a gene sequence to be found through a database of sequences that is homologous to a sequence of known function (American Medical Association, 2000). Because this sequence with an unknown function may be inferred to act similarly to that of the known sequence based on the known similarities, patents have been issued based on probable future determined function (American Medical Association, 2000).

History of the HGP

The HGP began on October 1, 1990 as a joint effort between the US Department of Energy (DOE) and the National Institutes of Health (NIH) (Human Genome Project, 2000). Although the project was planned to be completed in 15 years, the expected completion date has been set for 2003 (Human Genome Project, 2000). The initial idea for this DNA analysis arose in the early 1980s where scientists at the Department of Energy's Los Alamos and Lawrence Livermore National Laboratories were trying to determine whether mutations in the DNA of offspring of survivors of World War II atomic bombs was a result of the parents being exposed to radiation (Human Genome Project, 2000). Several pilot programmes were initiated in 1997 (Human Genome Project, 2000). In 1988, Congress appropriated $10.7 million for the DOE (Human Genome Project, 2000). In addition, in 1988 the DOE and NIH signed a 'Memorandum of Understanding', which led to the formal induction of the HGP (Human Genome Project, 2000). The NIH, led by its National Human Genome Research Institute (NHGRI), conducts much of its work through a university based grant programme (Human Genome Project, 2000). The HGP also encompasses international partners which included the UK, France, Germany and Japan (Human Genome Project, 2000).

In 1998, in the race to sequence the human genome, two private firms, Celera Genomics of Rockville, Maryland, and Incyte Genomics, formerly known as Incyte Pharmaceuticals, entered the arena of genome sequencing (Human Genome Project, 2000). Earlier that year, Perkin–Elmer Corporation, now known as PE Corporation, of Norwalk, Connecticut, and Dr J. Craig Venter and The Institute for Genomic Research (TIGR) of Rockville, Maryland, announced the formation by Perkin–Elmer and Dr Venter of Celera Genomics, a genomic company, which would 'substantially complete the sequencing of the human genome in three years'. Such efforts would be done using 'breakthrough DNA analysis technology' (Human Genome Project, 2000). Incyte Pharmaceuticals, however, had a different directive. In August 1998, Incyte announced it would sequence 'all commercially relevant [human genome] information' in 12–24 months (Human Genome Project, 2000). Incyte, concerned with the commercial value of the project, does not intend to decode the entire genome, comprising of some 3 billion units of DNA (Human Genome Project, 2000). Instead, it intended to only decode about 90% (Human Genome Project, 2000). The public sector, in an effort to complete the project before Venter, announced they would release a 'rough draft' covering 90% of the genome in September 1998. Production of such a draft and making it public was, as admitted by Dr Francis Collins, currently Director of NHGRI, a strategic move to weaken any patent position Celera or other businesses might claim (Roberts, 2001).

In addition, an issue of contention between the private and public group was database access. One of the goals set out by the HGP is to store the sequenced genome in public databases, placing it in the public domain, which would have the effect of emasculating all patenting rights. In October 1999, Venter and his group at Celera had sequenced one million bases of the human genome (Roberts, 2001). Venter followed shortly by announcing that his group had compiled DNA sequences covering 90% of the genome (Roberts, 2001). The public consortium grew upset with Venter after he announced his group would release their data on the Celera web site and not into the public database, Genbank (Roberts, 2001). In an effort to get other private companies to follow suit and openly share data, on 14 March 2000, former President Bill Clinton and British Prime Minister Tony Blair agreed the two countries would share fundamental data on the human genome, stating, '[w]e applaud the decision by scientists working on the Human Genome Project to release raw fundamental information about the human DNA sequence and its variants rapidly into the public domain, and we commend other scientists around the world to adopt this policy' (Human Genome Project, 2001). The public consortium and Celera Genomics each published a working draft of the human genome simultaneously, although in separate journals in February 2001, thus bringing to a close the first phase of the project (Human Genome Project, 2001).

Basic molecular biology

To understand the issues raised in patent law with biological materials, it is important to understand the science that has ultimately led to the invention. Therefore, a key to understanding 'biotechnology' is comprehending the terminology.

DNA, deoxyribonucleic acid, is the genetic material for all cellular life forms (Brown, 1999). It determines the physical characteristics of all living organisms. DNA is a polymeric molecule of four mononucleotide units (Lewin, 1997). Nucleotides are the basic unit of a DNA molecule and consist of three parts: a sugar known as a deoxyribose, a phosphate compound and any one of the four bases – adenine, thymine, guanine or cytosine (Lewin, 1997). The individual nucleotides are linked together to form a polynucleotide that contains any combination of adenine, thymine, guanine or cytosine (Lewin,

1997). DNA exists as a double helix consisting of anti-parallel strands in which the nucleotide units are linked by a 5′–3′ phosphodiester bonds (Lewin, 1997), and in which the nitrogenous bases are paired with one another such that adenine base pairs with thymine, and cytosine base pairs with guanine (Lewin, 1997). A gene is a unit of inheritance (Lewin, 1997). Contained within a nucleotide sequence of a gene is the information to direct the synthesis of a protein with a specific amino acid sequence using enzymes found within the cell through a ribonucleic acid (RNA) intermediate (Lewin, 1997). Therefore, the human genome is the complete set of DNA found in virtually every cell of the human body (Vacchiano, 1999). DNA is organized into 23 pairs of chromosomes in a human cell, each chromosome composed of one continuous molecule of DNA arranged into functional segments of DNA (Vacchiano, 1999).

Recombinant DNA technology

Recombinant DNA or genetic engineering is a molecular technique involving the 'splicing together of DNA from different sources and placing the recombinant DNA code into another cell' (quoted in Karczewski, 2000, p. 1050). The biotech industry has used genetic engineering techniques to manufacture large amounts of a desired protein for important research and commercial biomedical products. In order to obtain a desired protein using recombinant DNA technology, the complementary DNA (cDNA) that encodes the protein must be cloned (Karczewski, 2000). This process, of cloning a gene comprises of three steps. They are: '(1) the selection of a DNA source for cloning; (2) the production of a series of DNA fragments that can be inserted into a plasmid vector and introduced into a bacteria host; and (3) the isolation of the protein through the screening of the cDNA library for the desired sequence' (Karczewski, 2000, p. 1050).

The biotechnology industry arose in the 1970s (Karczewski, 2000). Today, patents play an important role in providing incentives to support the role of commercialized biotechnology. Because significant capital for initial research and development are needed, gene patents give assurance to investors in genomic companies, that these companies will be able to profit from their research and development interests (*Gene Patents and Other Genomic Inventions, 2000*).

Expressed Sequence Tags (ESTs)

The patent system in recent years has changed from the issuance of patents on conventional drugs to patents encompassing biological molecules containing genetic information (Bobrow and Thomas, 2001). Such patents include genomic DNA sequences, cDNAs, individual mutations, ESTs and single nucleotide polymorphisms (SNPs) (Bobrow and Thomas, 2001).

An EST is a length of cDNA that is generally only a partial sequence of a gene being expressed at the time a specific tissue is sampled (Holman and Munzer, 2000). An individual EST is not individually selected for sequencing, but is randomly isolated with many other stretches of cDNA from a cDNA library or from the tissue itself (Holman and Munzer, 2000). Although ESTs do not specifically define functional genes of proteins, they are useful in providing information and isolating functional, full-length genes (Vacchiano, 1999). In addition, ESTs provide structural information regarding human genes, probes for discovering complete genes, and markers for specific locations on the human genome. Furthermore, ESTs provide an opportunity for rapidly and inexpensively determining the extent to which a gene is expressed in a given cell type or tissue (Vacchiano, 1999).

Despite their utility, ESTs have inherent limitations. In most cases, ESTs do not provide the complete sequence of a cDNA (Vacchiano, 1999). Since genes are composed in part of related or identical segments of DNA, a researcher sometimes cannot unequivocally determine the gene from which an EST was derived without further analysis (Vacchiano, 1999). Secondly, ESTs are limited in that the methods employed to obtain rapid sequence determinations are less accurate than the more typical sequencing methods (Vacchiano, 1999). Thirdly, the routine methods used in discovery of ESTs, although highly efficient, eliminate some of the creativity and inventiveness traditionally employed in gene discovery (Vacchiano, 1999).

ESTs initially became controversial back in the early 1990s, when they first appeared. In 1991, Dr Venter, then of the National Institute for Neurological Disorders and Stroke (Roberts, 2001), used methods that focused on obtaining partial nucleotide sequence information from cDNA (Vacchiano, 1999). He named these partial nucleotide sequences 'expressed sequence tags' or 'ESTs' (Vacchiano, 1999). Knowing that in order to claim precedence for international patent rights, the patent application would have to be filed before the information is published, Reid Adler, head of the NIH Office of Technology Transfer, filed an application on the first group of 347 ESTs before Venter's 1991 article appeared in *Science* (Davies, 2001). Adler's justification was the fear that other countries might seek patents on some ESTs, and NIH filing would pre-empt such a move. Adler defended NIH's policy by saying, '[o]ur primary goal is to get products developed. Having patent protection will enhance our ability to transfer this technology to companies. Without patent protection, companies won't spend the money to develop it' (Davies, 2001). A year later, before Venter's next publication, Adler modified the application by adding 2421 additional ESTs (Davies, 2001). The application now covered more than 2500 ESTs, representing as much as 5% of the total number of human genes (Davies, 2001). Both academic and industry groups placed NIH under harsh scrutiny, once the application became known to the public (Davies, 2001). James Watson, Director of the Genome Center, challenged the 'non-obviousness' of Venter's sequencing technique. Other objections centred on the utility of ESTs (Davies, 2001). The patent application claimed not only the partial sequences but also their full-length genes and proteins derived from those genes (Davies, 2001). Venter had no experimental data showing the function of the genes with which to claim utility (Davies, 2001).

In 1992, the USPTO rejected the first round of applications, ruling the claims failed to meet any of the three criteria – utility, novelty and non-obviousness (Davies, 2001). The USPTO stated the claims were 'vague, indefinite, mis-descriptive, incomplete, inaccurate and incomprehensible' (Davies, 2001). NIH's claim was also weakened because a 15-lettered segment had been identified in some of the ESTs in other genes in the database (Davies, 2001). Thus, one could argue that the invention was obvious because a researcher could have theoretically isolated the EST based on the presence of the 15-letter base sequence (Davies, 2001).

Five years later, the USPTO had a change of heart and the acting Deputy Commissioner of Patents and Trademarks, Lawrence Goffney, announced that the USPTO would allow ESTs to be patentable based on their utility as probes (Henahan, 2001). He cited precedent regarding the patentability of works that were useful only as tools, noting 'we give patents for screwdrivers blades with a new slope everyday' (Henahan, 2001).

A biotechnology company that has developed a particular process for isolating ESTs can now patent the process and the ESTs it finds, thus fulfilling the novelty requirement. The ESTs are considered non-obvious only in the lab and not in the human body. The utility requirement has been able to be overcome with the USPTO accepting ESTs being used to find other ESTs (Wertz, 2001). These patents tend to be narrowly drawn. This favours research because it limits the possibilities of interference (Wertz, 2001). Today, however, emphasis has shifted from patenting DNA fragments such as ESTs to patenting cDNA or the entire gene (Wertz, 2001). The patenting of either a process that involves isolation of DNA or DNA-associated materials such as RNA, as well as chemical substances related to DNA is referred to as 'gene patents' (American Medical Association, 2000).

Patentability of full-length genes vs. ESTs

Two issues which relate to the patentability of ESTs with respect to claim scope are: (i) claims relating to the full-length gene from which

ESTs are derived, and (ii) whether EST themselves should be patentable. Although the USPTO has said ESTs are patentable if it can be shown that they are useful, this issue still remains controversial. For most genomic companies, ESTs alone have little commercial importance (Holman and Munzer, 2000). For example, ESTs are themselves patentable subject matter. Therefore, if an inventor lays claim to only an EST and he/she fails to make a claim to a larger gene from which the EST is derived, then the patentee does not have much value in the EST (Holman and Munzer, 2000). This is the case particularly when you have companies obtaining patents on as many gene fragments as possible (Holman and Munzer, 2000). These companies find value in trying to patent the gene itself. However, for some genomic companies such as Genome Sciences and Incyte Pharmaceuticals, the value comes from the databases and not the ESTs themselves. It is the gene and its protein product which provides for commercial value, particularly in drug development and medicinal therapies (Holman and Munzer, 2000).

With genome sequencing, a typical molecular biology application is the isolation, cloning and expression of a gene, which includes a naturally occurring protein (Cooper, 2000). The protein itself cannot be patented because it is a naturally occurring product of nature. However, an inventor can still patent the modified or purified form, which is considered a non-naturally occurring analogue, or claim the protein as a composition, which comprises the protein and additional components (Cooper, 2000). A way of claiming a protein is by reciting its sequence. Because this can be quite lengthy, the USPTO has issued rules whereby the sequence must be set forth in a 'Sequence Listing'. The patent application will disclose the amino acid sequence and assign each sequence in the application with a SEQ ID NO, and then list the sequence in the Sequencing List (Faber, 2000). US Patent No. 6,180,364 is representative: '1. An isolated DNA molecule comprising a DNA sequence which encodes EPIL/placentin protein having sequence ID 2 or analogues thereof having EPIL/Placentin-type action and obtained by deletion and/or substitution.'

Also with respect to US Patent No. 6,180,364, is the term 'comprising'. Comprising in claim drafting is a transitional phrase which means 'the invention includes the components especially mentioned in the claim, but does not exclude others which are not mentioned in the claim' (Gordon and Cookfair, 1995, p. 87). With respect to claiming the full-length of the gene, such claims usually will not run into patentability problems, when the inventor has identified the function of the gene and determined a use for it (Holman and Munzer, 2000) and the full-length gene or an identified portion of it is disclosed either in the claim itself or in the specification (Gordon and Cookfair, 1995). The significance of claiming the full-length of the gene is that it enables a patent holder to control use of that gene. Hence, for ESTs, a claim to a whole gene entitles the patentee to control the use of the entire gene, although the EST is only a part of the gene (Holman and Munzer, 2000). However, such a claim is limited to the scope of the disclosure of the patent (Holman and Munzer, 2000).

The following claims demonstrate ways in which patent protection relating to a protein may be obtained.

US Patent No. 5,621,080

1. An isolated erythropoietin glycoprotein having the *in vivo* biological activity of causing bone marrow cells to increase production of reticulocytes and red blood cells, wherein said erythropoietin glycoprotein comprises the mature erythropoietin amino acid sequence of FIG. 6 and has glycosylation which differs from that of human urinary erythropoietin.

US Patent No. 5,547,933

1. A non-naturally occurring erythropoietin glycoprotein product having the *in vivo* biological activity of causing bone marrow cells to increase production of reticulocytes and red blood cells and having glycosylation which differs from that of human urinary erythropoietin.

US Patent No. 6,291,644

1. A protein comprising an amino acid sequence SEQ ID NO. 1.

US Patent No. 6,291,235

1. Isolated nucleic acid molecule which encodes the protein portion of a protein-containing molecule which binds to monoclonal antibody A33, wherein the protein portion has a molecular weight of from 40 to 45 kDa as determined by SDS–PAGE under non-reducing conditions.

As shown above, the practice is that patents directed to the compositions of matter and/or distinct methods corresponding to these compositions can issue to more than one party. In a scenario where Inventor A has been granted a patent on the composition of protein X, and Inventor B has been granted a patent having claims directed to a new use of the composition of protein X, Inventor A may preclude others, including Inventor B, from making, using or selling protein X, whereas Inventor B may preclude others, including Inventor A, from using the composition of protein X according to Inventor B's patented use (*Primer: Genome and Genetic Research*, 2001).

The written-description requirement effect on patent litigation

Beyond an invention being useful and novel, a patent application must adequately disclose a description of the invention. The written-description requirement of 35 U.S.C., Section 112, first paragraph, sets forth the requirements for the description portion of patent applications. This section requires that the specification describe three things: the invention, the manner of making and using the invention known as enablement, and best mode (35 U.S.C.A. §112 Rejections Not Based on Prior Art, 2001). The requirement is met 'if the application contains sufficient disclosure, expressly or inherently, to make it clear to persons skilled in the art that [the applicant] possessed the subject matter claimed' (*Ex parte Harvey*, 1987). The requirement arose from the patent policy that the scope of an inventor's rights should not be as broad as his original invention (Holman and Munzer, 2000). It ensures that an inventor does not claim concepts or inventions that are in the public domain or belong to other inventors, and tells the public the boundaries of the inventors' patent protection (Holman and Munzer, 2000). In 1991, with *Vas-Cath, Inc. vs. Mahurkar*, the Federal Circuit failed to provide a clear standard for the written-description requirement by stating that an applicant 'does not have to describe exactly the subject matter claim' (*Vas-Cath, Inc. vs. Mahurkar* quoting *Rengo Co. vs. Molins Mach. Co.*, 1991, p. 1563).

Recent decisions made by the Court of Appeals for the Federal Circuit have indicated that to meet the written-description requirement for a claim to a gene the specification must recite the complete nucleotide sequence (Vacchiano, 1999). In *Amgen, Inc. vs. Chugai Pharmaceutical Co.*, the US Court of Appeals for the Federal Circuit took steps to clarify what the written description involved. Recall, Amgen had sued Chugai Pharmaceuticals for infringing a claim to a DNA sequence 'consisting of essentially of a DNA sequence encoding human erythropoietin' (*Amgen, Inc. vs. Chugai Pharmaceutical Co.*, 1991). The court ruled it was not enough to know how a compound of unknown structure, in this case the EPO gene, might be isolated to claim conception (Van Brunt, 2000). Instead, the inventor must actually isolate the gene (Van Brunt, 2000). In the USA, an invention is conceived 'when the party conceived the idea of the complete and operable invention' (Hildreth, 2000). According to the Federal Circuit, 'it is not sufficient to define [a gene] solely by its principal biological property, for example, encoding human erythropoietin…When an inventor is unable to envision the detailed construction of a gene so as to distinguish it from other materials, as well as a method for obtaining it, conception has not been achieved until reduction to practice has occurred or until after the gene has been isolated' (Van Brunt, 2000). This ruling has the effect of preventing individuals or companies from getting very broad coverage on a claim on all DNA sequences that code for a protein or analogue of that protein (Van Brunt, 2000).

This issue was revisited in 1993 in *Fiers vs. Revel and Tiollais vs. Sugano, Maramatsu and Taniguichi* (commonly referred to as *Fiers vs. Revel*). *Fiers* dealt with an interference proceeding between Fiers, Sugano and Revel, for priority of invention to the DNA sequence that encodes human fibroblast beta-interferon (1993). The patent application of Sugano disclosed the entire nucleotide sequence, while Revel's disclosed only a partial sequence and Fiers relied on conception (*Amgen, Inc. vs. Chugai Pharmaceutical Co.*, 1991) and reduction to practice to obtain the sequence (*Fiers vs. Revel*, 1993). The US PTO's Board of Patent Appeals and Interference awarded priority to Sugano, basing its conclusion on the disclosure of the complete nucleotide sequence of the DNA encoding the β-IF. The Court of Appeals agreed and concluded Sugano was 'entitled to rely on his disclosure as enabling (35 U.S.C.A. §112, 1984) since it sets forth a detailed teaching of a method for obtaining a DNA coding for β-IF…' (*Fiers vs. Revel*, 1993). This ruling demonstrates that one cannot claim a protein by demonstrating how to find the sequence, as relied on by Fiers; rather, one must have the sequence.

There have since been several other cases that have centred on the written-description requirement. In *The Regents of the University of California vs. Eli Lilly and Co.* (1997), the court, in determining the validity of two patents, considered the written-description requirement. The University of California claimed that Lilly had infringed its patents in its manufacturing of human insulin (1997). One of UC's patents was based on the determination of cDNA sequences from rats, but the claims covered mammalian and human insulin cDNA (*The Regents of the University of California vs. Eli Lilly and Co.*, 1997). The court held that in order for a DNA invention to meet the written-description requirement, it 'required a kind of specificity usually achieved by means of the recitation of the sequence of nucleotides that make up the cDNA' (*Regents of the University of California vs. Eli Lilly & Co.*, 1997). The court, citing *Fiers vs. Revel* stated, 'an adequate written description of a DNA requires more than a mere statement that it is part of the invention and reference to a potential method for isolating it; what is required is a description of the DNA itself' (*Regents of the University of California vs. Eli Lilly & Co.*, 1997, pp. 1566–1567 citing *Fiers vs. Revel*, 1993, p. 1170). Although the specification was enabling, the court concluded that an adequate description of the invention '"requires a precise definition, such as by structure, formula, chemical name, or physical properties," not a mere wish or plan for obtaining the claimed chemical invention' (*Regents of the University of California vs. Eli Lilly & Co.*, 1997, p. 1566).

In the case of ESTs, nothing is known about their structure or function, beyond the tissue sample from which the EST was derived. Opponents of EST patents could argue that ESTs fail the written-description requirement because they fail to shed light on the full-length sequence of the gene. They do not enable one ordinarily skilled in the art of molecular biology to know anything about the gene sequence. ESTs are only used as probes. Such an argument is most similar to *Fiers vs. Revel*, which taught that if one does not have the entire sequence of the gene, one does not have possession of the gene. The Court in *Eli Lilly* followed such a rationale.

Ethics

The USPTO has stated repeatedly that patents do not confer ownership of genes. The patent system was designed to 'promote progress' (Tebo, 2001). The following is a list of some potential arguments in favour of patenting biotechnological inventions:

1. Researchers are awarded for their discoveries.
2. Investment of resources is encouraged because a patent provides a 'monopoly' to the inventor by prohibiting competitors from making, using, selling the invention without a licence.
3. Duplication of effort is prevented if the system is used correctly by researchers.
4. Research is required to enter into new and unexplored areas.
5. Secrecy is reduced and thus all have access (http://www.ornl.gov/hgmis/elsi/patents.html).

In contrast, the following is a list of some potential arguments against patenting biotechnological inventions:

1. Patents of partial and uncharacterized cDNA sequences will reward those making routine discoveries and penalize those determining biological function or application.
2. Allowing a single genomic sequence to be patented in several ways such as an EST, a gene and a SNP may discourage product development because of high royalty costs owed to all patent owners of that sequence.
3. Private biotech companies who own certain patents can monopolize gene tests.
4. Patent filings are replacing journal articles and thus reducing the knowledge available (http://www.ornl.gov/hgmis/elsi/patents.html).

Species alteration (be it animal models or plant genetics) cannot be stopped as long as there is life. The question becomes not whether it will occur but whether it should occur under regulatory controls? If so, who should do the controlling? This is certainly not the role of the USPTO. The PTO was established as 'an office in the Department of Commerce, where records, books, drawings, specifications, and other papers and things pertaining to patents and to trademark registrations shall be kept and preserved, except as otherwise provided by law' (35 U.S.C.A. §1, 2001). Nowhere under 35 U.S.C.A. does it state that the USPTO is responsible for regulatory controls of what should be patented. It may be the role of other regulatory agencies such as the FDA, the USDA and possibly the states under their reserved police powers. The legislative history of the Transgenic Animal Patent Reform Act reports: 'We have not asked – nor should we ask the Patent Office – to act as a health and safety regulatory agency. Concerns about animal welfare can be – and are – addressed in a separate regulatory framework' (1989). However, the patent system itself should not be tinkered with to accomplish what Congress and the respective state legislatures should specifically address in the proper exercise of their powers and duties. 'Science is responsible, if man is' (Sease, 1988a, p. 572).

Conclusions

'Our Patent system has served us faithfully since the first patent law in 1970' (Sease, 1988a, p. 572). The landmark decision in *Diamond vs. Chakrabarty* will not change the course of rapid evolution of biotechnology (1979). The pioneering era of biotechnology will continue to evolve with the new guidelines of 2001. In addition, the Federal Circuit's strong pro-patent positions currently taken with biotechnological inventions will likely continue the current trend. 'To paraphrase Abraham Lincoln: the fuel of incentive provided by the patent system should continue to be added to the fire of biotechnology genius' (Sease, 1988b, p. 572).

References

35 U.S.C.A. §1 (West Supp. 2001).
35 U.S.C.A. §101 (1984).
35 U.S.C.A. §101 Part III (1984).
35 U.S.C.A. §102 (1984).
35 U.S.C.A. §102(a) (1984) (stating that '[a] person shall be entitled to a patent unless the invention was known or used by others in this country, or patented or described in a printed publication in this or a foreign country, before the invention thereof by the applicant for patent').
35 U.S.C.A. §103 (West Supp. 2001).
35 U.S.C.A. §103(a) (West Supp. 2001).
35 U.S.C.A. §103(b)(1) (West Supp. 2001).
35 U.S.C.A. §112 (1984).
35 U.S.C. §112(1) (1984).
35 U.S.C.A. §112 *Rejections Not Based on Prior Art*, <http://www.uspto.gov/web/offices/pac/dapp/oppd/35usc112.htm> [Accessed 4 October 2001].
35 U.S.C.A. §154(a)(2) (West Supp. 2001).
1077 *Official Gazette U.S. Patent & Trademark Office* 24 (21 April 1987).
Act of 21 February 1793, §1, 1 Stat. 319.

American Medical Association, *Gene Patenting*, <*http://www.ama-assn.org/ama/pub/printcat/2314.html*> [Accessed 22 August 2000].

Amgen, Inc. vs. Chugai Pharmaceutical Co. (1991) 927 F.2d 1200–1219.

Amgen, Inc. vs. Chugai Pharmaceutical Co. (1991) 927 F.2d 1200, 1206. The Federal Circuit in *Amgen* stated: 'Conception does not occur unless one has a mental picture of the structure of the chemical, or is able to define it by its method of preparation, its physical or chemical properties, or whatever characteristics sufficiently distinguish it. It is not sufficient to define it solely by its principal biological property, e.g., encoding human erythropoietin, because an alleged conception having no more specificity than that is simply a with to know the identity of any material with that biological property.'

Application of Bergy (C.C.P.A. 1979) 596 F.2d 952–1002.

Bobrow, M. and Thomas, S. (2001) Patents in a genetic age. *409 Nature*, Macmillan, London, pp. 763–764.

Brenner vs. Manson, (1966) 383 U.S. 519–540.

Brody, cf. Baruch (August 1991). Should we allow the patenting of life? *Agricultural Bioethics F.*, at 4 (arguing for patent rights as just desserts). For John Locke's original discussion, see John Locke, Second Treatise of Government 8–30, 1690 (C.B. Macpherson (ed.) Hackett, 1980).

Brown, T.A. (1999) The rule of DNA-binding proteins. In: *Genomes*. Wiley-Liss, New York, pp. 146–169.

Burchfiel, K.J. (1995) *Biotechnology and the Federal Circuit* 58, note 77.

Burchfiel, K.J. (Supp. 1997) *Biotechnology and the Federal Circuit* 10.

Burroughs Wellcome Co. vs. Barr Laboratories, Inc. (1994) 40 F.3d at 1223–1234.

Cooper, I.P. (2000) Claim Drafting. In: *Biotechnology and the Law*. West Group, Minnesota, pp. 5B-1–5B-83.

Davies, K. (2001) *Cracking the Genome.*The Free Press, New York, pp. 61–63.

Diamond vs. Chakrabarty (1980) 447 U.S. 303–322.

Division of Extramural Research, *The Human Genome Project* <http://www.nhgri.nih.gov/HGP/> [Accessed 15 September 2001].

Eisenberg, R.S. (1990) Patenting the human genome. *39 Emory L.J.* 721–745.

Enzo Biochem, Inc. vs. Calgene, Inc. (1999) 188 F.3d 1362–1381.

Ernst & Young, LLP (2001) *Focus on Fundamentals: The Biotechnology Report* (one of the world's leading professional services organizations which helps companies identify and capitalize on business opportunities), pp. 1–14.

Erramouspe, M. (1996) Comment: staking patent claims on the human blueprint: rewards and rent-dissipating races. *43 UCLA L. Rev.* 961–998.

Ex parte Deuel (1993) 27 U.S.P.Q.2d 1360–1365, *overruled by In re Deuel* (1995) 51 F.3d 1552, 1560 (while overruling the Board on the obviousness issue, the Federal Circuit left open further review by the PTO on other statutory grounds, including utility).

Ex parte Drulard (1983) 223 U.S.P.Q. 364–366 (holding patentable a portable lightening rod which, even though potentially unsafe, was sufficiently useful to satisfy 35 U.S.C. §100's utility standard).

Ex parte Harvey (1987) 3 U.S.P.Q.2d 1626–1630.

Ex parte Maizel (1992) 27 U.S.P.Q.2d 1662–1671.

Faber, R.C. (2000) Claiming biotechnology inventions. In: *Landis on Mechanics of Patent Claim Drafting*, 4th edn. Practising Law Institute, New York, pp. IX-1–IX-36.

Feisee, L. and Stanton, B. Are biotechnology patents important?, *PTO Pulse*, *<http://www.uspto.gov/go/opa/pulse/200003.htm>* [Accessed 24 September 2001].

Fiers vs. Revel (1993) 984 F.2d 1164–1172.

Funk Brothers Seed Co. vs. Kalo Innoculant Co., (1948) 333 U.S. 127–138 (holding that certain strains of each species of root-nodule bacteria are ‘no more than the discovery of some of the handiwork of nature and hence [are] not patentable’).

Gene Patents and Other Genomic Inventions: Hearing Before the Subcommittee on Courts and Intellectual Property of the Committee on the Judiciary House of Representatives (2000) 106th Congress 63.

Genetech. Inc. vs. Novo Nordisk A/S (1997) 108 F.3d 1361–1368 (quoting *In re Wright* (1993) 999 F.2d at 1557, 1561).

Genetic Face-Off, *<http://www.abcnews.go.com/sections/living/DaileyNew/genepatent 000228.html>* [Accessed 12 October 2001].

Gordon, T.T. and Cookfair, A.S. (1995) Patents as legal documents. In: *Patent Fundamentals for Scientists and Engineers*. CRC Lewis Publishers, Boca Raton, Florida, pp. 85–90.

Graham vs. John Deere Co. (1966) 383 U.S. 1–37.

Hearing Charter For Hearing on The Human Genome Project: Subcommittee on Energy and Environment, Committee on Science. U.S. House of Representatives (2000).

Henahan, S. *Patent Decision Shocks Genome Researchers*, *<http://www.accessexcellence. com/WN/SU/09/est297.html>* [Accessed 24 September 2001].

Hildreth, R.B. (2000) Interference. In: *Patent Law,* 3rd edn. Practising Law Institute, New York. pp. 8-1–8-9.

Hoffert, S.P. (6 July 1998) USPTO issues biotech patent guidelines. *The Scientist 12(14)* (arguing that narrow biotech-specific patent law principles may, ultimately, have the effect of chilling the development of new biotechnological inventions), available from: <http://www.the-scientist.com/yR1998/july/ hoffert_p1_980706.html [Accessed 8 October 2001].

Holman, M.A. and Munzer, S.R. (2000) Intellectual Property rights in genes and gene fragments: a registration solution for expressed sequence tags. *85 Iowa L. Rev.* 735–845.

Human Genome Project Information <http://www.ornl.gov/TechResources/ Human_Genome/home.html> [Accessed 27 August 2001].

Human Genome Project, Microsoft® Encarta® Online Encyclopedia *<http://www.encarta.msn.com/fin/print.sap?&pg=8&ti=0AC83000&sc=1&pt=1* > [Accessed 15 September 2001].

Hybritech, Inc. vs. Monoclonal Antibodies, Inc. (1986) 802 F.2d 1367–1385.

In re Bergy (C.C.P.A. 1977) 563 F.2d 1031–1042.

In re Chakrabarty (1979) 444 U.S. 924.

In re Deuel (1995) 51 F.3d 1552–1560.

In re Fisher (C.C.P.A. 1970) 427 F.2d 833–840.

In re Ruschig (C.C.P.A. 1967) 379 F.2d 990–996.

Iwasaka, R.M.T. (2000) Chakrabarty to chimeras: the growing need for evolutionary biology in patent law, *109 Yale L.J.* 1505–1534.

Karczewski, L.A. (2000) Biotechnological gene patent applications: the implications of the USPTO written description requirement guidelines on the biotechnology industry. *31 McGeorge L. Rev.* 1043–1086.

Kight, A.T. (1998) Note: pregnant with ambiguity: credibility and the PTO utility guidelines in light of breeder. *73 IND. L.J.* 997–1024.

Lewin, B. (1997) *Genes VI*. Oxford University Press, New York, pp. 51, 76–77, 80 and Fig. 4.11.

McGraw-Hill Encyclopedia of Science and Technology, 7th edn, 1992, p. 648.

Michaels, C.A. (1994) Biotechnology and the requirement for utility in patent law. *76 J. pat. & Trademark Off. Soc*, 247–260.

Miller, C.J. (1997) Patent law and human genomics. *26 Cap. U. L. REV*. 893–922.

Mueller, J.M. (1998) The evolving application of the written description requirement to biotechnological inventions. *13 Berkeley Tech. L.J.* 615–652. Available from: <http://www.ornl.gov/hgmis/elsi/patents.html>.

O'Reilly vs. Morse (1853) 56 U.S. 62–137; (holding natural principles or discoveries to be unpatentable); *Ex parte Latimer*, 1889 Dec. Comm'n Patent 123 (1889) (holding plants to be unpatentable).

Olsen, B.V. (1997) The biotechnology balancing act: patents for gene fragments, and licensing the 'Useful Arts'. 7 *ALB. L.J. SCI. & TECH.* 295–334 (citing *Marsa, L.* (December 1995) Whose ideas are they anyway: intellectual property in the information age). *OMNI* at 36 (quoting Bruce A. Lehman, United States Commissioner of Patents and Trademarks).

Primer: Genome and Genetic Research, Patent Protection and 21st Century Medicine, *<http://www.bio.org/genomics/primer.html>* [Accessed 8 October 2001].

Regalado, A. (September–October 2000). The Great Gene Grab, *MIT's Magazine of Innovation Technology Review*, pp. 50–55.

Regents of the University of California vs. Eli Lilly & Co. (1997) 119 F.3d 1559–1575.

Regents of the University of California vs. Eli Lilly & Co. (1997) 119 F.3d 1556–1557 (citing *Fiers vs. Revel* (1993) 984 F.2d at 1164, 1170).

Report of the President's Commission on the Patent System 1–3 (1966).

Roberts, L. Controversial from the start. *Science Online <http://www.sciencemag.org/cgi/contents/full/291/5507/1182a>* [Accessed 10 October 2001].

Sampson, M. (2000) The evolution of the enablement and written description requirements under 35 U.S.C. §112 in the area of biotechnology. *15 Berkeley Tech. L.J.* 1233–1274.

Scalise, D.G. and Nugent, D. (1995) International intellectual property protections for living matter: biotechnology, multinational conventions and the expectation for agriculture. *27 Case W. Res. J. Int'l L.* 83–118.

Sease, E.J. (1988a) From microbes, to corn seeds, to oysters, to mice: patentability of new life forms, *38 Drake L. Rev.* 551–572.

Sease, E.J. (1988b) *38 Drake L. Rev.* at 572. (The author first saw these words over the door of a US PTO in Washington, DC. This language comes from a speech delivered by Abraham Lincoln in Springfield, Illinois, in 1859).

Singer, P. (1990) *Animal Liberation*, 2nd edn. The New York Review of Books, Inc., New York.

Stewart, M.J. (1999) Note: The Written Description Requirement of 35 U.S.C. §112(1): The Standard After Regents of the University of California vs. Eli Lilly & Co. *32 Ind. L. Rev.* 537–564.

Tebo, M.G. (April 2001) The big gene profit machine. *87 ABA Journal*, 46–52.

Transgenic Animal Patent Reform Act (1989) H.R. 1556, 101st Congress.

US Congress, Office of Technology Assessment (1988) *Mapping Our Genes – The Genome Projects: How Vast, How Fast?* U.S. Government Printing Office, Washington, DC, p. 46.

US Constitution art. I, §8, cl. 8.
US Department of Commerce (1992) Business Forecasts for 350 Industries. *U.S. Industrial Outlook* '92, 17–1.
US Patent No. 4,736,866 (Issued 12 April 1988).
US Patent No. 5,190,931 (Issued 2 March 1993).
US Patent No. 5,547,933 (Issued 20 August 1996).
US Patent No. 5,621,080 (Issued 15 April 1988).
US Patent No. 6,180,364 (Issued 30 January 2001).
US Patent No. 6,291,235 (Issued 18 September 2001).
US Patent No. 6,291,644 (Issued 18 September 2001).
United States Patent and Trademark Office (21 January 2000) *Revised Interim Utility Guidelines Training Materials.*
Utility Examination Guidelines (14 July 1995) 60 Fed. Reg. 135, pp. 36263–36265.
Utility Examination Guidelines (5 January 2001) 66 Fed. Reg. 4, pp. 1092–1099.
Vacchiano, E. (1999) It's a wonderful genome: the written-description requirement protects the human genome from overly-broad patents. *32 J. Marshall L. Rev.* 805–832.
Van Brunt, J. Biotech patent fights. *Signals Magazine <http://www.signalsmag. com/signalsmag.nsf/657b06742b5748e888256570005cba01/3366125331d79f678825695b0058efaf?OpenDocument>* [Accessed 15 September 2000].
Vas-Cath, Inc. vs. Mahurkar (1991) 935 F.2d 1555–1569.
Vas-Cath, Inc. vs. Mahurkar (1991) at 1563 (quoting *Rengo Co. vs. Molins Mach. Co.* (1981) 657 F.2d at 535, 551.
Wertz, DC *Gene Letter-Patenting DNA: a Primer. <http://www.geneletter.com/archives/dna.html>* [Accessed 22 August 2001].

Open Source and Other Software Intellectual Property Models

9

Scott Newman[1] and Bruce L. Golden[2*]

[1] *PIC USA, PO Box 348, Franklin, KY 42135, USA;*
[2] *Optibrand, LLC*

Introduction

> We have to push this open source business. Else everyone is reinventing the wheel, and that is currently being done at a staggering rate. (Eildert Groeneveld, 2000, personal communication)

It is only since 1996 that software became patentable under US law. The guidelines were developed to assist patent examiners to establish the patentability of both hardware and software. The number of patents awarded for software and data-processing-related functions has climbed steadily since that time (see below). The main reason for patenting software and allied processes is the strength of protection provided by the patent laws to prevent the making, using or selling of the patented invention.

Contrary to this view is the community of open source developers whose main goal is to eliminate the barriers on copying, redistributing, understanding and modifying computer programs and processes. The open source movement had its beginnings in the early 1990s but became larger and more recognizable in 1994, when version 1.0 of the Linux operating system kernel was released. Other open source products in common use today include Apache, Perl and Netscape.

The objective of this chapter is to compare and contrast patent and open source issues. The benefits, as well as the shortcomings and challenges of open source will be elucidated. Economic and value-added models for the open source framework are discussed, as well as remedies for registering software and software-related inventions without the use of patenting.

*Current address: Department of Animal Sciences, Colorado State University, Fort Collins, CO 80523, USA.

Software Patenting

Computer programs have been under development for many years. However, the patentability of software did not become an issue until 1972 (*Gottschalk vs. Benson*), where the US Supreme Court attempted to decide whether the algorithm proposed in the patent as a 'process' was covered within the meaning of the patent act. The court ruled that the claimed invention did not fit within the subject matter of inventions as a 'process' but did not rule out ever finding a computer program patentable. Further controversy and specific rulings germane to this discussion can be found, for example, in writings by Kuester *et al.* (1997). The US Patent and Trademark Office (PTO) officially established guidelines for 'computer implemented inventions' in 1996.

Software as a class is not patentable *per se*. The guidelines developed by the PTO define three areas that are patentable: a machine that is directed by a computer program; a computer-readable memory or a computer-implemented process. The PTO considers four areas not to be patentable: compilation or arrangement of data, independent of physical element; memory encoded with creative or artistic expression; data structures; and an algorithm that manipulates only abstract ideas, e.g. an equation. For example, US PTO class 717 (software development, installation and management)

> …Provides for software program development tools and techniques including processes and apparatus for controlling data processing operations pertaining to the development and maintenance of software programs. Such processes and apparatus include:
>
> **1.** Processes and apparatus for program development functions such as specification, design, generation, and version management of source code programs.
> **2.** Processes and apparatus for debugging of computer program including monitoring, simulation, emulation, and profiling of software programs.
> **3.** Processes and apparatus for translating or compiling programs from a high-level representation to an intermediate code representation and finally into an object or machine code representation, including linking, and optimizing the program for subsequent execution.
> **4.** Processes and apparatus for updating, installing, and version management of developed code.

The number of patents awarded since 1996 under classification 717 includes: 157 patents awarded in 2000, 493 (2001) and 20 in 2002 (as of January 2002). So, although the patent classification has been available since 1996, it was only in 2000 that patents were actually being awarded. We expanded our search to include the 700 series classification [which includes all patents awarded pertaining to data processing (e.g. software development, database and file management, artificial intelligence, generic control systems, modelling and simulation, electrical computers and digital processing, and error detection)]. The results of this are presented in Fig. 9.1. While only 72 were awarded by 1997, over 2000 were awarded in 1998. There were over 9500 awarded in 2001, but the rate seemed to be slowing. About 750 patents were awarded in January 2002.

Given the listing of what can and cannot be patented previously described, it is best to examine the spectrum of software-related inventions to get an idea of their patentability. The ends of the spectrum are clear, but the lines in the middle are fuzzy (Kuester *et al.*, 1997). The categories include:

1. Mathematical algorithms found not to be patentable by themselves:
  a. Software algorithm for converting from binary-coded decimal numerals to pure binary numerals (*Gottschalk vs. Benson*, 1972).
  b. An improved mathematical equation for interpreting seismic waves (*In re Walter*, 1980).

2. Algorithms expressed as mathematical formulae found not to be patentable by themselves, and those algorithms which appear to replicate human decision-making skills found not to be patentable:
  a. A process for determining 'abnormalities' in a complex system (*In re Grams*, 1982).

3. Computer algorithms unrelated to mathematics found to be patentable:
  a. Computer program for the translation of natural languages (*In re Toma*, 1978).

4. Computer algorithms pertaining to the operations of the hardware found to be patentable:

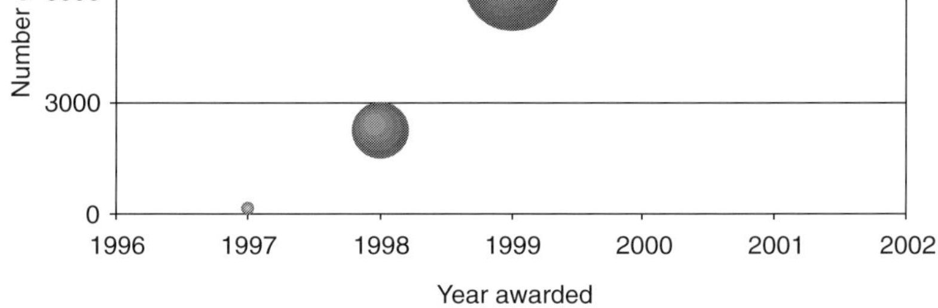

Fig. 9.1. Number of patents awarded under class 700 by US PTO since 1996.

a. System software (Application of Chatfield, 1976).
b. Algorithm for typesetting alphanumeric equations (*In re Freeman*, 1978).
c. A memory with data structure for organizing data (*In re Lowry*, 1994).

5. Processes or apparatus that use computer programs as a component of the overall invention found to be patentable:
a. Apparatus for voice pattern recognition (*In re Iwahashi*, 1989).

Besides the protection offered by patents preventing the making, using or selling of the patented invention, patents also provide companies the opportunity for cross-licensing its own patents (when accused of infringing a patent belong to another company), and creating a more effective defence to patent infringement suits (Kuester *et al.*, 1997).

Open Source Software

Open source software is software for which the source code is distributed or accessible via the Internet without charge or limitations on modifications and future distribution by third parties. Open source is a certification mark owned by the Open Source Initiative (OSI). The complete text of the Open Source Definition can be found at http://www.opensource.org/osd.html. The most common of the Open Source Licences is the Free Software Foundation's (GNU) General Public Licence (GPL). The various forms of the GPL can be found at http://www.gnu.org

Open source is the result of a long-time movement toward software that is developed and improved by a group of volunteers cooperating together on a network. Several flavours of the Unix clone operating systems are open source, including Linux and FreeBSD. The Open Source Definition is an effort to provide a branded model or guideline for this kind of software distribution and redistribution.

The precursor model to open source was developed by Eric Raymond (1999) and is called the Cathedral and Bazaar (CatB). He discussed open source as two fundamentally different development styles (see Fig. 9.2). The 'cathedral' model is the case in which very few developers labour in isolation to produce a piece of software, and the 'bazaar' model is the case in which software is created with a large, loosely affiliated group of developers with limited central control. The Cathedral represents a monolithic, highly planned, top-down, more centralized style of software development. The Bazaar represents a chaotic, evolutionary, market-driven model (Raymond, 1999).

Fig. 9.2. A depiction of the Cathedral and Bazaar model of open source development. © 1998, Illuminata Inc., Nashua, New Hampshire, USA. Artist: John Libby.

Open Source and Traditional Methods of Software Development

There are a myriad of methods available for engineering software (e.g. Pressman, 2000), from linear designs to more evolutionary and iterative processes. The issues here are not so much the processes of developing, writing and evaluating software as much as they are about the economics of software development. However, the distribution model chosen can radically influence the development method used.

As discussed earlier the CatB model is a metaphorical reference to two fundamentally different styles of software engineering. The Cathedral represents the centralized planning model, where top-down development is the norm and implementation is by special project teams around structured schedules. Efficiency is of primary importance here (the bottom line). Thus, the 'usual' model is about development of proprietary software where access to the code is restricted and customers are reliant on the company for upgrades and maintenance (bug fixes).

Alternatively, the Bazaar expresses a decentralized and bottom-up approach to development driven by volunteer hackers.[1] It appears to be chaotic and unstructured, a community where no one alone is effectively in charge.

[1] Here a hacker is a clever programmer. It is not someone who cracks someone else's system or otherwise uses programming or expert knowledge to act maliciously. The latter are referred to as 'crackers' by the software development community (Raymond, 1999).

The following points summarize the differences in what open source software provides as opposed to the usual model of software development and distribution:

- the open source software being distributed must be redistributed to anyone else without any restriction;
- the source code must be made available without cost (so that the receiving party will be able to improve or modify it);
- the licence can require improved versions of the software to carry a different name or version from the original software; and
- improvements cannot be made proprietary and must be distributed under the open source software agreement.

What are the strengths and weaknesses of open source?

The perceived strengths and weaknesses of the open source model have been sourced from a number of web references, including Mantarov (1999), Raymond (1999) and Varner (1999).

Perceived strengths of open source

- Allows (and even encourages) many people to share enhancements and detect fixes (parallel debugging) to software (collectively called 'modifications'), which leads to better quality, higher reliability, lower costs and increased choice;
- allows independent peer review on a scale much larger than before by an independent community;
- growth of, and access to the Internet allows communication from groups all over the world;
- people will contribute new features to increase the program's value to them (parallel development);
- allows for early and frequent releases for bug detection;
- treats users as co-developers and accepting their contributions is the 'least-hassle route to rapid code improvement and effective debugging' (Raymond, 1999);
- open source allows for risk sharing;
- enhances communication with downstream developers;
- provide patterns and examples for software methods that can be applied to other applications;
- promotes the development and adoption of standards (e.g. OpenSSL);
- valued software is supported, but bad is not.

Perceived weaknesses of open source

- Assumes that open source is good and that commercial development is bad – no continuum of developmental models;
- open source software development is compared against only two exemplars (the Cathedral and the Bazaar). However, this model does not account for the size of the project, its complexity, time-frame and time pressures, and availability of resources/tools;
- open source initiatives can become extinct if the leader of a given project overworks and burns out, if there is an inability to acquire a critical mass of users, if there is loss of a leading developer, or if 'forking' occurs (starting your own 'mutant' version of any open source project);
- lack of customer focus;
- no entity is accountable;
- no centralized technical support;
- no effective economic model to make money from development.

World Wide Web resources have become available in recent years to provide repositories for open source code and allied applications. One such portal is SourceForge.net (http://www.sourceforge.net), which provides project hosting, version control, bug and issue tracking and project management tools.

The Economic Value of Software

Computer programmes have three primary types of value (Varner, 1999); their economic value as a tool (use value), their market value as a saleable commodity, and the value of the software by virtue of not only owning the programme, but also having it unavailable to competitors (monopoly value). By the sheer nature of open source, the market value and the monopoly value of the software is removed, and the use value is increased by virtue of the product being more stable and robust.

However, open source makes it difficult to capture direct sale value from software. This has more to do with the social contracts of open source developers than technical difficulties, since the source code is available (Raymond, 1999). These social contracts have to do with extraction of profits by one party, the perceived 'chilling' effect of licences and fees for commercial use, and the preservation of peer-review and maintaining the 'gift-culture' of open source (gift culture is an anthropological term in which members compete for status by giving things away). Thus, the Open Source Definition has the effect of making sale value very hard to capture. However, models do exist to develop markets in software-related services that capture at least indirect sale value. The two most successful models include loss leaders and service sellers.

Loss-leader/market positioner

In this model, open source is used to create or maintain a market position for proprietary software that generates a direct revenue stream, for example sales of server software or subscription/advertising revenue associated with a web portal. One example of the use of this model was the decision by Netscape to provide its browser code to accelerate development and debugging of the browser to gain market share from Microsoft (see http://home.netscape.com/browsers/future/whitepaper.html, 1998).

Service seller (give away the recipe, open a restaurant)

In this model software is open-sourced to create a market position for services (not software). This is where value-adding is in the assembly and testing and running an operating system that is warranted to be merchantable and to be plug-in-compatible with other operating systems carrying the same brand, and can also include free installation support and options for continuing support. An example of this model would be the Red Hat Linux distribution (http://www.redhat.com). This model has greatly contributed to the increased use of Linux as an operating system alternative.

Two other important economic models for open source include:

Widget frosting

A model for hardware manufacturers where they are forced to write and maintain software (e.g. device drivers and configuration tools), but the software itself is an overhead, not a profit center. There is no revenue stream to lose, and the vendor gains from a larger developer pool and better reliability. An example is Apple's decision to open-source the core of their 'MacOS X' server operating system (http://www.apple.com/pr/library/1999/mar/16opensource.html)

Accessorizing

Sale of accessories for open source software, including anything from mugs and tee shirts to professionally edited and produced documentation (see for example http://www.oreilly.com).

Three other, more speculative models include:[2]

[2] See www.ora.com/catalog/cb/chapter/magic-cauldron.html for a more complete discussion of these models.

Free the future, sell the present

Release software in binaries and source with a closed licence, which includes an expiration date on the closure provisions (e.g. permit free distribution and forbid commercial use without fee).

Free the software, sell the brand

Open-source a software technology, retain a test suite or set of compatibility criteria, then sell users a brand certifying that their implementation of the technology is compatible with all others wearing that brand.

Free the software, sell the content

Open-source all the software and sell subscriptions to the content. Then enhance software as new platforms are developed, automatically expanding your market.

In an open source model, value is not derived directly from the software. A possible exception to this is when the software is developed under a grant. Much of the open source software used in genetics/bioinformatics is developed this way. See for example http://www.sanger.ac.uk/Software or http://www.tigr.org/softlab. In many ways the development of bioinformatics software is a good example of the Bazaar approach to software development. The laboratory receives money to pay its staff and the organization/institution receives money as overhead. There are many examples of open source bioinformatics software then becoming a part of an integrated package and being redistributed, either in a wrapper or as a plugin. Generally, there seems to be little resistance to these efforts among the genetics research community. Partially, this is due to the fact that many of the commercial packages that leverage the open source software as a component of their package add value to the scientists by making it more feature rich or enhancing the ergonomics.

What Experiences are there in the Animal Breeding and Genetics Community?

Programs written by quantitative geneticists for various livestock industries can generally be categorized into data analysis and prediction tools, decision support tools and teaching tools (there is some overlap between the latter two). Examples of the three classes can be found in the appendix of Chapter 15. Programs defined as having source code available are open source in the sense that you can 'look at the source code'.

In the animal breeding and genetics community, developers of the code (where code is available) do not actively communicate through a 'community' of users and hackers to make improvements to the code, peer review and debug *per se*. Indeed, there seems to be a wide variety of ways in which code is made available for enhancements by others, as well as services for closed source applications. For example:

- BREEDPLAN (http://breedplan.une.edu.au), LAMBPLAN (http://lambplan.une.edu.au), Total Genetic Resource Management (http://tgrm.une.edu.au) and BreedObject (http://www.breedobject.com) require 'expert intervention' for their use. Thus they can all be considered as fee-for-service applications and source code is not freely available. These are examples of bureau- or service-oriented applications;
- LAMBPLAN uses Pedigree Wizard (http://www.lambplan.com.au/download/pwupdate.htm) as a database tool. The Pedigree Wizard code is not open source, but it has frequent and unplanned updates based on end-user suggestions;
- PIGBLUP (http://abri.une.edu.au/pigblup.htm) users pay a licence fee. However, the package is run from a desktop machine (no bureau service provided), and yearly upgrade fees provide for free workshops to teach users new modules in a continuing education process. Source code is not freely available;
- abtk, the animal breeder's toolkit (http://ops.agsci.colostate.edu/~bgolden/abtk) is true open source. It is a series of UNIX commands that complete various data manipulation and analysis processes in the estimation of breeding values for genetic evaluation. A number of modules have been developed and debugged by various users from around the world. These enhancements are made freely available to all end-users;
- VCE (http://www.tzv.fal.de/institut/genetik/vce4/vce4.html), MTDFREML and MTGSAM (http://aipl.arsusda.gov/curtvt/mtdfreml.html) and DFREML (http://agbu.une.edu.au/~kmeyer/dfreml.html) are based on what might be termed 'monolithic code' that would require a great deal of intervention on the part of the original developer to make the OS environment work. One developer invites others to his lab for further development of the package. Another stated:

> I have made both source code and compiled versions of [programme name] available for downloading via the [programme name] home page. No charge, but I expect kudos. Obviously that's not enforceable. No formal backup. I'll look at interesting queries, but replies to questions which can be answered by reading the manuscript or a little bit of trial and error tend to be terse to non-existent. I am not particularly interested in any tinkering of the code by others, other than people reporting blatant bugs, and I steadfastly refuse to look at results/problems/etc. from versions that have been modified by others. (Anonymous developer)

This was the general view of other developers as well. The source code is available but the original developers do not necessarily support changes. This does provide opportunity for bug detection and development of code for new models, but there is no process for including new options into already-existing code and distributing the newer package. Obviously many of these developers are delivering a free product to the animal breeding and genetics community, but they are not necessarily interested in enhancements made by others. Many of the developers queried were also concerned about recognition.

Who will maintain the effort (of programs whose code is freely available) if there is no financial incentive or if the developer's organization desires them to be spending more time on what they were hired to do? Often too many people want too many features – in this regard open source would be ideal. For developers, there is a very sharp double-edged sword: people will find errors and may be discourteous to the developer (bothered to fix bugs and how to use program). However, the original work can be greatly expanded through open source.

Another aspect of development is ownership based on funding for software development. Many of the products described above, especially those that are bureau- or service-oriented, have been developed with industry funds based on known statistical and genetic methods of analysis. The package of programs is marketed for a cost to the user. There is some return to the universities or commercial entities for the use of that package, but likely not in proportion to the funds required for configuration, delivery and maintenance. Thus, the commercializers make some profit from providing an integrated service to industry, and industry makes improvement through the purchase of genetically improved seedstock.

Alternatively, the Agricultural Research Service of the US Department of Agriculture (and other US Government agencies) hold that since a piece of software (like the Decision Evaluator for the Cattle Industry or MTDFREML, described above) evolved from public money, then the code should be accessible to others including commercial entities. No payment is expected. If a commercial entity changed a few lines of code and copyright that change, there is no objection. Transfer of enhancements made to the code to improve performance is welcomed by the developers, but there is no expectation that will occur (it is not mandatory). However, it would influence future working relationships if the sharing were not mutual.

It is interesting to note that there has been no impetus for developers of software in animal breeding and genetics to develop an open source-based community. However, there is a substantial base and tradition of freely distributed applications software for analysis of animal breeding data. There are probably many reasons why animal breeders have not embraced a more formal mechanism for protecting these applications using open source licensing. Probably the most significant is that animal breeding software has such a limited market that there is not much threat of coopting the application into a proprietary environment. Many animal breeders use their software as a way to communicate their ideas to colleagues, and a formal process of open source protection would be a burden with a perceived limited value. Finally, some animal breeders perceive that institutional administrators are reluctant to allow them to distribute software freely. Often, this is viewed as giving away something that may have revenue-generating potential. Most institutional administrators do not understand the limited size of the market for these applications. In order to avoid having the software bottled-up in a debate over the merits of seeking open source licensing, that would require administrative intervention, breeders simply release software into the public domain, unprotected.

Opportunities for developing networks and communities of developers for software development are required and would likely be welcomed if leadership were provided. However, until granting agencies require it from animal breeders, as they do for many bioinformatics development efforts, animal breeders will continue to not protect their freely distributed code.

Bioinformatics and Open Source

An enormous global infrastructure has been developed around bioinformatics since the 1980s. The decrease in the cost of capturing genome information (e.g. expressed sequence tags and single nucleotide polymorphisms) has caused a great increase in the sheer size of datasets available to molecular biologists, in spite of the fact that most sequence data is still held privately. Thus, the need to handle this data computationally has increased as well. It has been estimated that the sales of desktop bioinformatics programs in the USA will rise from $390 million in 2002 to $2.2 billion in 2007 (*Red Herring*, 2002).

A number of for-profit commercial software companies have come into being to take advantage of this burgeoning niche. Some of these companies often rely on open source programs (e.g. Apache Web Server) as building blocks for products released under proprietary licences. Thus, open source allows these companies to price its software more competitively or even lower than similar products. This is a valid open source business model, and holds true when the components are freely available. When they are not, the components need to be built from scratch. A dilemma occurs when groups that have received public funding might have already developed these kernels, but the code and/or algorithms have already been licensed to private corporations, preventing free access.

The controversy revolves around the access of bioinformatics tools developed under the auspices of public grants. Some would argue that the fruits of research developed under publicly funded agreements should be made freely available, as a condition of funding, under an open source licence. The Bayh–Dole Act of 1980 (P.L. 96-517, The Patent and Trademark Law Amendments Act) allow universities to patent federally funded research results. Open source licences like the GNU GPL prohibit exclusive licensing of software patents, and thus can conflict with a universities right to license what it has legally patented. These issues have yet to be resolved.

It has been argued that an open source model for bioinformatics is a natural part of the scientific process (Stewart, 2002). First, open source follows well-known scientific principles of openly sharing and discussing results (code in this case) and to allow third-party confirmation of results. Secondly, in biological research, the data are of more importance than the computer programs – researchers receive greater advantage by accessing other datasets, something that can be achieved by sharing software tools. Finally, genomics will be best served through the creation of a common infrastructure that can best be done with open source tools, as open source ensures that the largest number have access to equipment and methods.

A number of organizations have been set up to deal with open source issues. These include Bioinformatics.org (http://www.bioinformatics.org) and the Open Bioinformatics Foundation (http://www.open-bio.org).

Remedies and Alternatives to Patenting and Open Source

Copyright and copyleft

As explained in Chapter 1, copyright is a separate class of protection from patents, and involves fewer formal procedures. Copyright arises automatically once an original effort has been started and some aspect of it has been fixed in a tangible medium of expression (Field, 2000). Registration of copyright is required only if legal action is warranted. Registration of copyright is inexpensive and is completed in the US through the US Copyright Office (http://www.loc.gov/copyright/). By copyrighting software, it becomes illegal to copy or distribute the software or its documentation without permission from the copyright holder. However, this means only the source code, or executable image is protected from unauthorized copying. It does not supply adequate protection of the ideas. Intellectual property such as features, function, or screen layout is not protected by copyright.

The open source community has developed the 'copyleft' method as a process for making a computer program free software and requiring all modified and extended versions of the program to be free software as well (http://www.gnu.org/copyleft/copyleft.html). In its simplest form, copyleft states that anyone who redistributes software, with or without changes, must pass along the freedom to further copy and change the software. This prevents individuals or groups from making changes to software and then turning the result into a proprietary product. Again, this does not protect the ideas in the software. However, this is not normally a problem for copyleft developers because one of the reasons to copyleft software is to demonstrate use patterns and procedures. One of the real benefits of copyleft source code is its ability to show other developers alternative patterns for accomplishing tasks.

Prior art

A recent article in the *New York Times* (Milstein, 2002) described technology developed by Plantronics, Inc. to reduce microphone noise. While the company was not ready to utilize the technology, they chose to post a description of the invention on IP.com (http://www.ip.com) to establish the invention's legal existence. This is known as prior art in patent law. This is where novelty and non-obviousness are judged against everything publicly known before the invention, as shown in earlier patents and other published material. This disclosure is also known as defensive publishing. Obviously competitors will gain access to the invention, but this risk is often outweighed by the difficulty prior art provides for someone else to receive a patent, including licence fees.

The Foresight Open Source Disclosure Project, developed by the Foresight Institute (http://www.foresight.org) in association with IP.com, is one way to help open source developers establish prior art for their developments. For a small fee, a document submitted to the disclosure project via IP.com becomes part of a database made available to patent officers from all over the world to help in conducting prior art searches before a patent is awarded.

To Open-source or Not?

Raymond (1999) posed the following question: 'What is being protected when source code is closed?' The economically rational answer to this question can take two forms: (i) to sell the package to other people; or (ii) deny its use to competitors. Of course, this assumes there is either a market worth developing or a competitive advantage. This is very rare in the limited markets for animal breeding applications.

In (i), above, sale value is being protected. For (ii), the answer is not as well defined. Suppose a bioinformatics package is open-sourced. This act increases its popularity and the benefits that occur from the hacker community by sharing the code are realized. However, a competitor is using the package and gets the benefit of it without paying the development cost, which in turn cuts into profits. Is this an argument against open source? It is an interesting question to ponder.

The decision to open-source can be self-motivated, to demonstrate a developer's real skills. The hacker could release exceptional code under an open source model, received accolades from the community, but better still is offered a 'high salary' in a company to be their Chief Information Officer. In this way the hacker can then go on to build new and improved tools that are not released to the public.

Raymond (1999) argues that *the real question is whether the gains from spreading the development load exceed the loss due to increased competition from the 'free ride'*. In this regard, companies might ignore the functional advantage of recruiting more development help, and also the (erroneous) view of including development costs as part of the answer. The development costs had to be paid anyway, so counting it, as a cost of open sourcing, might not provide an unbiased appraisal.

Another reason often cited for closing source code is the fear of disclosing a particularly special function that might reveal confidential aspects of a procedure. It could be argued that this is not an argument for closed source but bad design. The logic here is that business knowledge should not be expressed in code but rather in some sort of specification language implemented by the program's engine. Thus, the engine could be open-sourced without the worry of imparting proprietary information.

Discriminators that push toward open source (Raymond, 1999) include when:

- reliability/stability/scalability are critical;
- correctness of design and implementation cannot readily be verified by means other than independent peer review;
- the software is critical to the user's control of his/her business;
- Software establishes or enables a common computing and communications infrastructure;
- key methods (or functional equivalents of them) are part of the common engineering knowledge.

Of course, one of the very unique opportunities that exist in moving toward an open source model would be the increased development of synergies with regard to ideas that go into the code rather than just development of the source code itself.

References

Application of Chatfield, 545 F.2d 152 at note 5 (C.C.P.A., 1976).
Field, T.G., Jr (2000) Copyright for computer authors. http://www.fplc.edu/tfield/CopySof.htm (last accessed 10 March 2002).
Gottschalk vs. Benson (1972) Supreme Court of the US. 409 U.S. 63.
In re Freeman, 573 F.2d 1237 at 1246 (C.C.P.A., 1978).
In re Grams, 888 F.2d 835 (Fed. Cir., 1982).
In re Iwahashi, 888 F.2d 1370 (Fed. Cir., 1989).
In re Lowry, 32 U.S.P.Q.2d 1031 (Fed. Cir. 1994).
In re Toma, 575 F.2d 872 at note 4 (C.C.P.A., 1978).
In re Walter, 618 F.2d 758, 205 U.S.P.Q. 397 (C.C.P.A., 1980).
Kuester, J.R., Moceyunas, A. and Jennings III, B.B. (1997) Patents for software-related inventions. http://www.tkhr.com/articles/pat4sof.html (last accessed 4 March 2002).
Mantarov, B. (1999) Open source software as a new business model. MSc thesis, The University of Reading, Reading.
Milstein, L. (2002) Protecting intellectual property. New York Times, 18 February http://www.nytimes.com/2002/02/18/technology/ebusiness/18NECO.html (last accessed 4 March 2002).
Pressman, R.S. (2000) *Software Engineering, a Practitioner's Approach*, 5th edn. McGraw-Hill Series in Computer Science, Place, Columbus, Ohio.
Raymond, E.S. (1999) *The Cathedral and the Bazaar, 1st edn.* O'Reilly, Cambridge.
Red Herring (2002) The open source debate enters the genomics arena, pp. 22–23 (March).
Stewart, B. (2002) Ewan Birney's keynote: a case for open source bioinformatics. The O'Reilly Network. http://www.oreillynet.com/pub/a/network/2002/01/28/bioday1.html (last accessed 4 March 2002).
Varner, P.E. (1999) The economics of open source. http://www.cs.virginia.edu/~pev5b/writing/econ_oss/resources.html (last accessed 4 March, 2002).

Animals, Ethics and Patents

10

Donald Evans

University of Otago Medical School, Dunedin Hospital, Dunedin, New Zealand

> Animals, who we have made our slaves, we do not like to consider our equal. (Charles Darwin, Notebooks 1837–1838)

Introduction

Some of the ethical dimensions of patenting genes, either human, plant or animal, are inextricably bound up with the ethics of the activity of identifying and modifying genes at all. These activities have produced more public debate about scientific research than any other, leading to numerous calls for moratoria on research of certain kinds in both humans, animals and plants (Evans, 1999, p. 109). Given that the patenting of gene sequences of one kind or another depends upon these activities, we cannot ignore the ethical questions which they provoke in exploring the ethical dimensions of gene patenting itself. I propose to canvass these problems in the first part of the chapter before going on to discuss subsequent ethical issues in patenting the outcomes of such research.

Ethics in Animal Research

The use of animals in research is both commonplace and controversial. The general public is probably unaware of the extent of such use despite the voluble protests of the animal rights movement. Discussions on the topic between the major protagonists tend to engender more heat than light and they have, in addition, been accompanied by threats and intimidation. Let us attempt to identify the central issues regarding the ethics of animal research in general before looking at the area of genetic research in particular.

A decent answer to the question 'What is good research?' will contain at least two elements. First the researcher will need to be skilful. Skills are means–ends activities that involve setting objectives, designing procedures to meet those objectives and executing those procedures. The most skilful researcher will, in design and execution, maximize outcomes whilst minimizing inputs, i.e. he/she will be maximally efficient. There are already ethical dimensions to consider in evaluating this aspect of research protocols. Inefficient practice, whether located in poor research design or execution, involves waste of resources. In times of fiscal stringency, resources wasted in one piece of research are denied to other worthy research projects. More significantly, perhaps, where animals are employed, the unnecessary suffering or destruction of living creatures as could occur, for example, where the numbers required were not accurately assessed, cannot be condoned.

Our answer would be seriously deficient if the ethical evaluation stopped at this point. The worthiness of the research goal as well as its realizability calls for evaluation. Both the investment of resources and the suffering and sacrifice of living creatures to worthless research are morally suspect. The capricious exposure of animals to harm or slaughter evokes moral concern in us all. The ethical evaluation of all research involving living subjects, including clinical research, involves the weighing of potential harms and benefits, and this might be regarded as an art rather than a skill.

Evaluation of research on human participants goes beyond a consideration of the skill factor, as does the sub-task of setting worthy research objectives. We need to consider how the research impacts on respect for a number of values that inform our relationships with other human beings. The values that figure most prominently are autonomy, beneficence, truthfulness, dignity, non-maleficence and justice. It is difficult to see how some of these values can be seen to figure in our relationships with animals, but it is not difficult to see how others can. As a matter of fact there is considerable dispute about where the line is to be drawn here and it is to that issue which I now turn.

The Ethical Domain of Animals

The instrumental use of animals in research can easily blind us to ethical dimensions of our relationships with them. This is consonant with other instrumental uses of animals as Darwin points out in the motto of this chapter. But what does he mean by his ironic use of the word 'equal'? Clearly there are some aspects of our relationship in which we can never be equal. We can, for example, have moral perceptions of them but they cannot have such perceptions of us, nor can they enter into a wide range of the spectrum of emotions, which we can experience, due to the fact that they are not language users and do not participate in the host of social institutions which identify our feelings and us. However, to jump from this fact to the conclusion that they are not morally considerable and deserving of our moral respect is to jump too far, even though it might be a convenient move to make. Where we draw the line will itself be a matter of moral dispute. The polarized views which would either demonize animal researchers or ridicule protesters as sentimentalists are each wide of the mark, but where in between can we find ground that respects the research enterprise and animals?

Darwin's use of the notion of slavery offers us a clue to a partial solution of this problem. Slaves were people who had become the tools of others. Their own independent lives counted for nothing, as their sole purpose was to effect the designs and wishes of those who had the right to own and employ them. That is not to say that all slave owners were sadistic and heartless exploiters of these lives. Far from it, for we have good evidence to distinguish caring and beneficent owners from others. Whether out of concern to preserve the most effective workforce or out of higher motives, many owners were willing to house, feed and protect their slaves. Nevertheless, slaves they remained, with no possibility of leading lives that were their own. In retrospect, we do not condone even the kindliest employment of slaves. However, within that tradition, the chances of an imaginative relationship between slave and master, which respected the independence of the life of the slave, were almost nil. The idea of respecting the independent life of a slave would not occur to an owner, for whom moral obligation ended with the decent treatment of the slave as a slave. The context of the relationship constrained the possibility of a wider sense of concern, which would have characterized the relationship between a slave master and an equal.

It is this constraint to which Darwin refers. We can see how the analogy fits by considering an example from animal research (Diamond, 1991, p. 354). Sir Lauder Brunton, a celebrated pharmacologist, was keen to demonstrate the humane character of his research on animals. He cited the example of a dog on which he had made a gastric fistula.

> The dog showed no sign of pain and, moreover, whenever the dog was examined it showed great delight – just like a dog that has been sitting about the house, and wants to run out for a walk. When it saw that I was going to look into its stomach, it frisked about in the same way as if I was going to take it out for a walk (Paget, 1912, p. 90).

Diamond observes that there is another possible reaction to this scene, namely 'what a miserable life for an animal' as opposed to 'oh good – no pain!' Thinking outside the square of the laboratory setting, Brunton would no doubt have deplored such restrictions in a dog's life which the mere absence of pain would not have mitigated. It is, she observes, for this reason that the charge that animal experimentation makes one callous is levelled. It is a case of special pleading which might result in carelessness in calculating the moral cost of such experimentation. That is not to say that, given an awareness of that cost, given a greater moral sensitivity on the part of the researcher, the research would not proceed. Some might not but some might, though with a sense of regret that such a price is demanded by the worthiness of the goal to be achieved.

If we need further stimulation to think of how easy it is to engage in the compartmentalization of mind to which Diamond refers, we have only to think back to the beginnings of serious ethical review of clinical research. The Nuremberg Code arose out of the Nuremberg trials in which Nazi doctors were indicted for crimes committed against humanity in the name of medical research. Their attitude to their research subjects was mirrored in the Nazi attitude to slave labour where, for political purposes, people were no longer seen as leading worthwhile individual lives. Some modicum of ethical regard was due them in this political square but its expression might strike us as an inversion of moral consciousness:

> We shall never be rough and heartless when it is not necessary, that is clear. We Germans, who are the only people in the world who have a decent attitude towards animals, will also assume a decent attitude towards these human animals (Himmler, 1943).

As surely as human beings in a given context can be seen in an ethically restricted frame, as human animals, so animals themselves, in the context of the laboratory, might be viewed in an ethically less imaginative light. They may become laboratory animals, animals to which a whole range of ethically significant attitudes is no longer appropriate. In genetic research, they might be regarded simply as repositories of gene sequences or test beds for genetic manipulation.

The Sentimentalist Criticism

A common reaction to protests against the use of animals in research is that to object to such practice is to sentimentalize the human/animals relation. How much purchase is there in such allegations?

How human beings relate to animals is in part facilitated by the manner in which animals enter into the lives of people. This is not universally the same in either time or place. People of different cultures at different times have developed a variety of interactions with animals which have resulted in a multitude of perceptions about their significance. These range from the worship of animals to their instrumental use for either food or labour. At the more fundamental level we are able to ascribe certain psychological descriptions to animals because of certain similarities between themselves and us. What we are prepared to say will vary from animal to animal.

> It is easy enough to extend the concepts of 'sensuous' experience to creatures fairly like human beings, such as cats, dogs, and horses; when we try to extend them to creatures extremely unlike human beings in their style of life, we feel, if we are wise, great uncertainty – not just uncertainty as to the facts, or as to the possibility of finding them out, but uncertainty as to the *meaning* of saying: 'I now know how to tell when an earthworm is angry' (Geach, 1957, pp. 113–114).

The possession of a face is particularly important in this regard, which is why some vegetarians refuse to eat anything with a face. This has little to do with sentimentalizing animals. Just try to doubt that a snarling dog baring its teeth is not angry!

However, animals can enter into the lives of people at a more profound ethical level than this. It is here that moral obligation to animals takes on a deeper meaning than simply not causing suffering or distress. Prince Llewellyn is said never to have smiled after thrusting a spear in to the heart of his hound Gelert. He had mistaken the blood upon its jowls for the blood of his infant son, whom the Prince had left in the dog's protection, only to discover subsequently the dead form of a wolf lying beside the unharmed child. Gelert is a byword for faithfulness in Welsh culture and Llewellyn suffered guilt for his lack of trust of the dog. We could enumerate many examples of sophisticated relationships of this sort between man and animals. The pit ponies who laboured in the Welsh coal mines easing the labour of the miners were rewarded with summer holidays in the fresh air and peaceful retirements in the countryside. Why should this be regarded as sentimentality rather than the repayment of a debt?

It would therefore be a mark of moral insensitivity to regard animals in research settings merely as sophisticated mechanical instruments, as has been the case (Eccles, 1971, pp. 285–293).

Animal Rights

The most extreme form of opposition to the use of animals in research comes from those who claim that animals have rights which demand our respect. This includes the right to be protected from harm resulting from coercive treatment. The consequence of the extreme forms of this view is that no research employing animals should proceed as they are not able to consent to such procedures, given that they are incapable of comprehending the requisite information (Regan, 1984, pp. 381–382).

The most prominent academic proponents of this view take the view that we should behave towards living beings, including human beings, in ways which take due account of their interests which are dependent on their capacity for suffering and enjoyment. The result is that some animals will be seen to have interests which some human beings cannot have because of their relative capacities. For example, it would be mistaken to be more concerned for the welfare of a senile or insane human being than for a healthy higher primate. This mistake would be evidence of a prejudice against animals which has been labelled 'speciesism' (Ryder, 1975).

There are numerous problems associated with this view. Not least amongst them are the counter-intuitive conclusions which follow from it, such as the one mentioned above. Such conclusions might be construed as constituting a *reductio ad absurdam* of the position from which they are derived, rather than a compelling case for changing moral attitudes towards human beings and animals. It has been argued that to confuse the question of the differences between animals and humans (such as their relative capacities in different circumstances) with the question of the difference between human beings and animals, a pervading difference in attitudes and conceptualization arising out of the fabric of human life and institutions, affords the position more plausibility than it deserves (Diamond, 1991, p. 324). The moral difference between humans and animals is not based on our response to their various capacities but rather on our perceptions of what they are. It is this difference which explains why we do not eat human beings and why we conduct funerals at their death, not their capacities to feel distress or pain. As we have seen above, animals can engage in aspects of our lives which call for sophisticated ethical responses, but never in such a way as to erase this difference, but that is because of our relationship with them and not because of innate capacities which they have.

> Most people would of course think that if, during... an expedition, a situation arose in which a dog's interests and a person's clashed (say, only one could be rescued from a crevasse) the dog's would properly be sacrificed. But many people would *also* think that at the end of the expedition the dogs should not be killed (however painlessly) merely to save the money it would cost to ship them back (Diamond, 1991, p. 340).

On this view the use of animals in research to save human lives would be permissible, whereas their use to produce cosmetics would be contestable. There could be other justifications outside medical applications, but these would each need to be assessed on a case-by-case basis. One way in which valuable research might be conducted on animals, whilst maximizing respect for them, would be to resist rushing into the target animal in the research process by first exploiting other laboratory processes such as the use of tissue culture. Both the publication of research on human subjects in pharmaceutical development and the licensing of products developed in such research are dependent on the research procedures having been subjected to independent ethical review. If the research is thought to be unethical, then it should not find its way into learned papers and the outcomes should not find their way into commercial activities. Such restrictions on patenting inventions in animal research would help to ensure that due respect to animals was safeguarded.

Species Integrity

Engagement in genetic research raises additional ethical problems. The use of animals in this field is subject to these concerns. The problems of risk to and the safety of humans and the environment loom large in the debates, in addition to concern for the welfare of the animals involved (Evans, 1999, pp. 109–111). Thus, for example, the great hopes invested in xenotransplantation have been dented by the fear of retroviruses, in days when uncontrollable diseases such as AIDS and BSE have been attributed to viruses and proteins crossing species boundaries. Similarly, worries about control and the difficulties of containment cause protests about genetic experimentation on plants and animals.

I wish to consider a further, and perhaps more fundamental, objection to these procedures. It is an ideological objection which marks off animal genetic research from other research on animals. Asserting intellectual property rights (IPRs) to life forms is thought by some to encroach on what is sacred or taboo. Similarly, interfering with species boundaries is thought to exhibit hubris. For example, in their submission to the Royal Commission on Genetic Modification the Catholic Bishops said:

> We believe that all human beings have a role to play as co-creators with God, and as participants in the evolutionary process. The responsible use of GM is a further challenge for us as stewards of creation … Because we participate with God in the continuing evolution of life in all its forms; we also have a responsibility to respect the integrity of other species and all creation (Nathaniel Centre, 2000).

Similar affirmations of respect for species inform the Maori submissions.

What can researchers make of such objections? The Catholic Bishops' statement seems to provide both support and a fatal objection to the transfer of genes between species. It is difficult to see secure, given boundaries between species if one takes the evolutionary process seriously. Yet respect for the integrity of species appears to demand such security of knowledge. Perhaps the claim really collapses into an appeal to nature, which would be in line with the Catholic position on other matters such as interference in the human reproductive process. Insofar as this is true, then the submission has much in common with the Maori submissions. Discussions have proceeded on the plausibility of such an appeal but they do not provide conclusive reasons for or against the view (Pickering, 1996, pp. 111–125). We are thus left with a fundamental moral disagreement which is represented, for example, in the general views of the New Zealand population on genetic modification (IBAC, 2000, pp. 15–16). How public policy on genetic modification should be formulated in a pluralist society is perhaps the most difficult issue, and it spills over into gene patenting regulation. Those responsible for developing patent law are somewhat sensitive to this difficulty, as can be seen from the following statement made in European discussions of the matter:

> … that inventions shall be considered *unpatentable* where exploitation would be *contrary to public policy or morality* (European Commission, 1996).

The statement has in mind *inter alia* germline modification and the use of embryos for industrial or commercial purposes. We need to ask what is the character of such objections. They are not all religious in the formal sense, as the Maori objections would be labelled spiritual, in that they are not connected with formal doctrines but rather reflect the intimate relationships which people are thought to share with other life forms. They are not even all spiritual in the widest sense, as deep ecologists would also contend for the inherent value of species as being indisputble.

The dispute about species integrity is not empirical. No amount of evidence, new or old, can be guaranteed to move the protagonists on either side of the divide, though it might move some in either direction. It is tied up with the notion of awe or wonder. This is a feature of religious belief which is threatened by science whose nature it is to press for explanations of phenomena (Churchill, 1994, p. 405). Churchill claims that 20th-century thought has substituted the endless quest for explanation in terms of reasons for a willingness to stop when we have a solution. The demand for explanation is a recursive demand, by always seeking a justification that shows why this way is the way, and it functions to block another attitude towards the practice – a wonder or astonishment that it does go this way – happens to go this way – rather than another. Wonder can be seen as an attitude to things or as a form of attention which is undermined by explanation.

Churchill helps us with the analogy of Mark Twain's boyhood wonder at the variegated textures and colours on the surface of the Mississippi which was destroyed by his training as a steamboat pilot:

> The unsignifying glory of the River, its aesthetic beauty, was lost to him forever, and in its place he saw a system of signs, signals of clear passage, shallow water, cross currents, underwater snags (Churchill, 1994, p. 411).

So it is that the explanation of human life in genetic terms tends to dispel a certain kind of ignorance, but at the same time makes other kinds of attention to human and other forms of life more difficult to achieve. It is not that they have been shown to be invalid, in the sense that they are based somehow on an empirical mistake. Rather, life is seen consequently as a more mundane thing and is thus diminished by it. This in turn can undermine respect for both human beings and animals in research, making it more likely that things which are of fundamental value will be commodified (Hanson, 1997, p. 3). This is precisely what ideological critics claim occurs when life forms are patented. Thus, we end up knowing the cost of everything but the value of nothing.

Of course, it does not have to be like this. The scientist might be struck by the magnificent complexity of life as science sees it and, for example, see God's glory manifested in it, but if that is difficult for the scientist it is much more difficult for the non-scientist. The difficulties are compounded by the pseudo-religious language often employed by scientists to describe the character of their work where, for example, DNA is described as 'offering the key to the mystery of life' (Nelkin and Lindee, 1985, p. 39).

In the end, the decision has to be a political one, in the face of a lack of moral consensus, but this does not mean that it matters not what researchers do with animals, or indeed where they do it. The New Zealand example of the transgenic cows is a case in point, where the land on which they have been produced is Maori land and their presence is morally offensive.

There is a connected issue affecting indigenous peoples where there might not be an objection in principle to the procedures. It is that to claim IPRs to knowledge of their flora and fauna is to plunder their heritage. This has happened in pharmaceutical developments and is happening again, allegedly (Mies, 1993, pp. 238 and 242–244), in animal and plant genetics. There might be fundamental objections, from all the sources considered above, to what is seen as purloining for profit what is given us all by nature, but even where such objections are not made there are ethical concerns to be addressed. These concerns have led countries where such exploitation has occurred to neatly characterize the activity as biopiracy (Downes, 2000). It follows that some considerations of shared commercial interests between indigenous people and those who exploit their heritage are called for. Such provision has been made in principle in Articles 3 and 15 of the 1992 Convention of Biological Diversity, where the sovereign right of states over their genetic resources are asserted together with provisions for a fair and equitable sharing of the results of research and development, and the commercial benefits arising from the utilization of these resources.

Patenting Life Forms

The ethical misgivings about patenting life forms are most pronounced where whole animals are patented. As late as 1972, even microorganisms were not patentable matter under US law because they were deemed to be 'products of nature'. On appeal it was agreed that patent law did not cover 'living things', but the Court of Customs and Patent Appeals ruled 'that the fact that microorganisms are alive is a distinction without significance for patent law', whereupon, on further appeal, Chief Justice Warren Burger held that 'a live, human made microorganism is patentable subject matter. [The inventor's] discovery is not nature's handiwork, but his own' (Deftos, 2000). This was extended on 7 April 1987 by the Patent and Trademark Office to cover 'nonnaturally occurring, nonhuman, multicellular living organisms, including animals'. A year later the transgenic oncomouse was patented. Human beings are not included in the remit of patenting law. However, Article 5(2) of the proposed European Directive (European Commission, 1996) maintained that while the human body and its elements in their natural state were not patentable inventions, an invention capable of industrial application which relates to an element isolated from the human body or otherwise produced by means of a technical process can be patented 'even if that element is identical to that of a natural element' (Elizade, 1998, p. 321). Elizade indicates an even longer history of patenting living things (1998, p. 318). There was a case in Finland as early as 1843, and Louis Pasteur patented a yeast free from organic germs of disease in 1873. The landmark judgement was the *Diamond vs. Chakrabarty* case in the USA in 1980, which declared that a genetically engineered microorganism is patentable subject matter.

Whilst genetic interference with life at a cellular level prompts moral objections, patenting animals *per se* has caused greater concern. Indeed, even though it is now the case that animals have been patented in Europe (1992) and Japan (1994), there is still much concern about the issue in Europe and undeveloped countries (Deftos, 2000). Indeed there remains one first-world country, Canada, which continues to disallow animal patent applications.

Opponents of such practices cannot but feel that their commodification criticisms are valid when they see shopping lists of transgenic animals, such as oncomice which are genetically engineered to have susceptibility to cancers of various kinds, distributed to facilitate invasive and destructive research. Whilst for some people this may be seen as a tragic necessity, for others it will remain a morally unacceptable trade. Of greatest concern are those who see it as neither and simply press on, in a blinkered manner, without any thought for the independent lives at stake.

The case for patenting genes and transgenic animals

There are important types of ethical arguments to be made for the patenting of genes and life forms. The first kind constitutes an appeal to fundamental values whilst the second emphasizes ethical consequences.

The principle of patenting has a longer history than the disputes we have considered thus far. As far back as the 17th century, the distinguished philosopher John Locke set out the principle of justice which underlies the practice:

> Whatsoever ... [man] removes out of the state that nature hath provided and left it in, he hath mixed his labour with, and joined to it something that is his own, and thereby makes it his property (John Locke, 1690).

The idea is that a man's labour should be properly rewarded. If that labour results in the production of something useful then he should have some power to determine who shall have the right to use it. Whilst patenting rights do not amount to property rights, there is considerable affinity between this general principle and patenting. It is a general idea whose appeal is enhanced by consequential considerations such as are found in the American Constitution, which gives Congress the power 'to promote the progress of science and useful arts, by securing for limited times to authors and inventors the exclusive right to their respective writings and discoveries' (U.S.C. Article 1, Section 8, Clause 8US). The criticism that for an individual to assume control of something which is given to us all amounts to theft is somewhat rebutted by the limited period of control that is granted to its discoverer, as well as by the fact that what eventually is given is endowed with the added value provided by the inventor. The dual view has been emphasized more recently by a prominent bioethicist:

> Everyone sees the at least *prima facie* claim that justice requires protecting the labours of the imaginative and industrious against theft by the sly and lazy ... The Foundation of the patent law is not only utilitarian, but also ethical (Kass, 1985, p. 135).

He goes on to say that it encourages virtues and rewards effort. Nevertheless, there is still room for moral disagreement when life forms are the subjects of patent. Both the grounds of justification have in fact been challenged in this field. We have already noted the claim of critics that patenting life forms amounts to improper appropriation of what is a common heritage. This criticism has been cast in the form of challenging the appeal to justice as a ground for issuing such patents (Burrows, 1997, pp. 11–14). Burrows' claim is that justice is infringed by patenting biological material from indigenous peoples to develop medical treatments because it claims something which rightfully belongs to others, and intimately so. The consequentialist justification is also challenged in the field of genetic patents. It is to this controversy which I finally turn.

Impediments to research and practice

The Constitution laid down that where 'the progress of science and the useful arts' would be facilitated by providing exclusive rights to the authors and discoverers then they should enjoy the same for a limited period. The advocates for patenting of life forms, including expressed sequence tags (ESTs), single nucleotide polymorphisms (SNPs) and transgenic animals, claim that advances in both science and the welfare of human beings in the provision of better health care and diet would be frustrated by a refusal to issue patents.

One of the major justifications of patenting human genes is that banning such activity could hold back advances in therapeutic research and practice (Reiss, 1996, pp. 15–19; Hanson, 1997, p. 3). This is certainly the view taken by industry following the example of pharmaceutical patenting where the profit motive, though condemned by many, nevertheless has ensured that many useful substances have been developed (Williamson, 2001). The dearth of useful products produced over 70 years in communist state-controlled countries is eloquent testimony to this.

However, there are dissonant voices to be heard on this subject. Take, for example the BRCA 1 gene, a marker gene for predisposition to breast cancer, for which a patent has been granted to Myriad Genetics, an American Company, by the European Patent office. The patent is being challenged by The Institut Curie, France's top centre for research diagnosis and treatment of cancer (Dorozynski, 2001, p. 589). The Institut claims that the patent 'gives Myriad an abusive monopoly and threatens basic research'. Doctors have to send their samples to Myriad and this amounts to 'an intolerable loss of data and expertise for French laboratories'. The test is also almost four times as expensive as a similar French test. The Institut further claims that the patent is too wide 'covering existing as well as future molecular screening techniques'. The challenge thus involves multiple objections to the claim that the patent facilitates research and therapy. Research and innovation could be stifled, patients could be denied tests and necessary treatments, and the development of research skills in France could be hindered. Similar disputes are ongoing concerning the patenting of the human embryonic stem cell in the USA by a foundation of the University of Wisconsin, and the proposed application by that foundation for a European patent.

Whether the latter application will be successful is a moot point given that two of the European Community's main policy platforms on genetics are the following:

> Intellectual property of inventions must allow for free circulation of data among researchers: patenting of isolated human genes is acceptable, as long as novel, inventive, useful products or processes are created, but such patents should respect ethical limits, allow for research purposes free use of genetic knowledge, and should exclude ownership or commerce of the human body and human materials.

> Equity of benefits of human genetic research: it should include access to health care based on biotechnology, but also allow transfer of these biotechnologies for organized access to genetic resources and respect of biological diversity (Elizade, 1998, pp. 320–321).

Of course, some safeguards are already built into patent law to avoid these unethical outcomes. For example, three criteria for patenting life forms have to be satisfied, namely that they are novel, useful and non-obvious. The 'novel' criterion means that they have not previously existed in description or in physical form (Murashige, 1994, p. 184) and the 'useful' criterion means that, for example, an EST is patentable if it provides a diagnostic test but not if it only invites the possibility of further research (Bonetta, 2001). This should deter the patent speculators who endeavour to patent ESTs and SNPs on the basis that someone, somewhere, might discover a function for the patented item and be obliged to pay royalties to the speculator for a considerable period following. Nevertheless, thousands of applications of this sort are pending (Flattman and Kaplan, 2000, p. 283), one company alone having filed applications covering fragments of 50,000 genes (Bonetta, 2001, p. 115), but this safeguard would not apply either to the BRCA 1 gene or to embryonic stem cells. Another safeguard would be to insist that patent applications not have the catchall nature which would cover other similar inventions. Yet advice is given to those who would frame patent applications that:

> If a company has not disclosed and claimed in its patents (or patent application) the broadest features of the invention – such as the varied characteristics of, or alternative designs of, the invention – then the patent protection obtained may not adequately prevent others from using either the invention or key aspects of it. This potentially opens the door to immediate direct competition (Praiss, 2001, p. BE6).

Patent offices should keep a keen eye on broadest characterizations if research and therapeutic applications are not to be stifled.

The threat to academic research

Academic research has traditionally been public good research. It is true that many good ideas produced by such research have floundered after publication in learned journals, for want of venture capital or the will or nous to develop them commercially. It is also true that freedom from the pressure to achieve short-term goals has facilitated major blue skies research. It is feared that the patenting of procedures or products essential to university research could have profound negative impacts on that research. The two most significant impacts are the move from basic to applied research and the unaffordability of university research.

The first is already happening on the principle of 'if you can't beat them then join them'. Universities are getting in on the act and encouraging research which will produce patentable outcomes and, in turn, handsome pay-offs. Whilst this will help avoid the non-development of good outcomes of academic research, it is also having a deleterious effect (Eisenstein and Resnick 2001, pp. 881–882). The worry is that university research policy could be distorted by the search for patents for applied knowledge rather than breakthrough knowledge. Given the relatively short life of a patent, the need to produce markets for a product and the time lag in obtaining licences, the big payers mostly concern methods rather than composition-of-matter. This is true of many pharmaceutical developments but is much more likely to be true of breakthrough genetic or stem cell patents. Universities have moved considerable attention to this kind of research given the possible returns offered. One example is the patent for recombinant DNA technology granted to Stanford University and the University of California, which earned $38.5 million in 1997–1998 alone.

The second impact is manifested in the fear expressed by some laboratory scientists that gene patents discourage them from developing diagnostic tests using patented genes for fear of being shut down by demands for payments (Bonetta, 2001, p. R116).

There is thus a balance to be struck between maintaining sufficient stimulus to facilitate useful research and innovation on the one hand, and protecting the interests of basic and public good research on the other hand. This will not be easily achieved but the price of failure will be high for both science and society.

Acknowledgement

I am grateful to the Association for the Advancement of Animal Breeding and Genetics (AAABG) for an invitation to think about some of the problems discussed in this chapter for a presentation at their Queenstown, New Zealand Conference, 2001.

References

Bonetta, L. (2001) Raising the bar for gene patents. *Current Biology* 11, R115-R116.

Burrows, B. (1997) Second thoughts about U.S. Patent #4,438,032. *Bulletin of Medical Ethics* 124, 11–14.

Churchill, J. (1994) Wonder and the end of explanation: Wittgenstein and religious sensibility. *Philosophical Investigations* 17, 388–416.

Darwin, C. (1987) Notebook B (1837–8) in P.H. Barrett *et al.* (eds) *Charles Darwin's Notebooks 1836–1844.*

Diamond v Chakrabarty (447 U.S.303) 1980.
Diamond, C. (1991) Experimenting on animals. In: *The Realistic Spirit. Wittgenstein, Philosophy and the Mind.* MIT Press, Boston, Massachusetts, p. 354.
Deftos, L.J. (2000) Patenting life: the Harvard mouse that has not roared. *The Scientist* 14, 6 November.
Dorozynski, A. (2001) France challenges patent for genetic screening of breast cancer. *BMJ* 323, 15.
Downes, D.R. (2000) How intellectual property could be a tool to protect traditional knowledge. *Columbia Journal of Environmental Law*, 25, 253–282.
Eccles, J.C. (1971) Animal experimentation versus human experimentation. In: *Defining the Laboratory Animal.* International Committee on Laboratory Animals and the Institute of Laboratory Animal Resources, National Research Council, National Academy of Sciences, Washington, DC.
Eisenstein, R.I. and Resnick, D.S. (2001) Going for the Big One. Nature Publishing Group, http://biotech.nature.com, Vol. 19, pp. 881–882.
Elizade, J. (1998) The patentability of human genes: an ethical debate in the European community. *The Journal of Medicine and Philosophy* 23, No. 3.
European Commission (1996) Proposal for a European Parliament and Council Directive on legal protection of biotechnological inventions. *Official Journal of the European Communities*, No. C296, 8.10.1996, Art 6(1).
Evans, D. (1999) Ethics and genetics. *The New Zealand Medical Journal* 112.
Flattman, G.J. and Kaplan, J.M. (2000) Patenting expressed sequence tags and single nucleotide polymorphisms. Nature Publishing Group, http://biotech.nature.com, Vol. 19, pp. 683–684.
Geach, P.T. (1957) *Mental Acts: Their Content and Their Objects.* Routledge and Kegan Paul, London.
Hanson, M.J. (1997) Religious voices in biotechnology: the case of patenting. *Hastings Center Report* 27, 1–19.
Himmler, H. (1943) Speech 4, October 1943.
IBAC, 2000. *Biotechnology in New Zealand, Consultation Report.* Independent Biotechnology Advisory Council, Wellington, New Zealand.
Kass, L. (1985) *Toward a More Natural Science: Biology and Human Affairs.* The Free Press, New York.
Locke, J. (1690) *Second Treatise of Civil Government*, Ch. 5, Sec. 27.
Mies, M. and Shiva, V. (1993) *vs. Ecofeminism.* Zed Books, Halifax.
Murashige, K.H. (1994) Intellectual property and genetic testing. In: Frankel, M.S. and Teich, A. (eds) *The Genetic Frontier: Ethics, Law and Policy.* American Association for the Advancement of Science, Washington, DC.
Nathaniel Centre (2000) *Nathaniel Report*, Issue No. 2, November.
Nelkin, D. and Lindee, S. (1985) *The DNA Mystique: the Gene as a Cultural Icon.* W.H. Freeman and Company, New York.
Paget, S. (1912) *For and Against Experiments on Animals: Evidence Before the Royal Commission on Vivisection.* New York, p. 90.
Pickering, N.J. (1996) Naturally conceived: the idea of the natural in moral arguments. In: Evans, D. (ed.) *Creating the Child: Ethics, Law and Practice in Assisted Procreation*, Martinus Nijhoff, Amsterdam.
Praiss, D.M. (2001) Creating a winning patent portfolio. Nature Publishing Group, http://biotech.nature.com, Vol. 19, pp. BE5-BE7.

Regan, T. (1984) *The Case for Animal Rights*. Routledge, London, 1984, pp. 381–382.
Reiss (1996) Is it ethical to patent human genes? In: Simmons and Simmons (eds) *Innovation from Nature. The Protection of Inventions in Biology*. London (http://k1.ioe.ac.uk/scitech/staff/mn/publications.doc; pp.15–19.)
Ryder, R.D. (1975) *The Victims of Science: the Use of Animals in Research*. Davis-Poynter.
Williamson, A.R. (2001) Gene patents: socially acceptable monopolies or an unnecessary hindrance to research? *Trends in Genetics*, in press.

11

Development of a Genetic Marker for Litter Size in the Pig: a Case Study

Max F. Rothschild[1] and Graham S. Plastow[2]

[1] *Department of Animal Science, Iowa State University, 2255 Kildee Hall, Ames, IA 50011, USA;* [2] *Sygen International, Fyfield Wick, Abingdon, Oxfordshire OX13 5NA, UK*

Introduction

Although reproductive performance is one of the most critical components in swine production, it has been one of the most difficult areas for geneticists and animal breeders to improve. Traditional quantitative methods such as selection, inbreeding and cross-breeding have had limited success in this area (Rothschild, 1996). This is primarily due to the low heritabilities, sex-limited nature and insufficient phenotypic measurements for reproductive traits. Consequently, considerable variation exists for traits such as litter size, which currently ranges from approximately two to 40 pigs per litter, with average litter size ranging from nine to 16 depending on breed (Rothschild, 1996). Therefore, litter size is very well suited for approaches based on marker-assisted selection.

Early investigations into the genetic basis of reproductive traits centred on the identification of blood groups and protein polymorphisms and their association with litter size. The estimation of potential pleiotropic effects of known major genes was also investigated. First evidence of such associations were obtained by Jensen *et al.* (1968) and Rasmusen and Hagen (1973), who reported an association between the H locus on pig chromosome 6 and litter size with an unfavourable effect of the Ha allele. Kristjansson (1964) and Imlah (1970) reported an apparent effect of alleles at the transferrin locus (pig chromosome 13) on pig fertility and prolificacy. Unfortunately, the transferrin locus effect was not confirmed by other investigators (e.g. Jensen *et al.*, 1968; Huang and Rasmusen, 1982). Several other associations between blood group loci and reproductive traits have often led to contradictory results (e.g.

Kennedy *et al.*, 1973). During the 1980s, there was considerable investigation of the role of the major histocompatibility complex (MHC) called the swine leukocyte antigen (SLA) complex on male and female reproductive traits. In the female, the MHC has been associated with ovulation rate (Rothschild *et al.*, 1984; Conley *et al.*, 1988), litter size, number of pigs born alive and number weaned (reviewed in Vaiman *et al.*, 1988). For a complete review, see Warner and Rothschild (1991). These results suggested a role for single genes in controlling litter size but results were indefinite at best. The results were freely published but the industry as a whole seemed reluctant to use this information in their breeding programmes, since the results were unclear and there was a lack of knowledge of how to implement such results.

In the 1990s, three changes occurred which moved swine genetics research closer to a realization that commercialization and patenting was not only possible, but perhaps desirable. First and foremost was the advent of molecular biology that allowed an entirely different method of exploring the genetic differences for reproduction that existed within the domesticated pig. This allowed for the study of so-called 'candidate' genes that could be involved with biological processes such as reproduction and traits like litter size. This new approach was significantly more expensive than previous research and required larger investments than breeding companies had previously made for research. Second, biotechnology companies were becoming interested in animal agriculture and were willing to fund this high-risk/high-tech research. Thirdly, thanks to efforts by the USDA, University of Illinois, Iowa State University (ISU) and a number of pork producer's associations, Meishan pigs known for their exceptionally large litters were imported into the USA from China, providing exotic germplasm for new research. The scientific curiosity, the financial interest and the genetic material to better understand reproduction were now in position to attack the problem of improving reproduction through modern genetic methods.

Estrogen Receptor (ESR) – an Interesting Candidate Gene

In the 1980s, Rothschild and colleagues had begun to investigate the role of the MHC on pig chromosome 7 and reproduction. Rudimentary comparative maps (O'Brien, 1993) were beginning to be published that made it clear that the MHC complex was on human chromosome 6 but at this time the breakpoints of the comparative map were not known. In late 1990, with the Meishan pigs at Iowa State University, this team began to be interested in linked genes that could control reproduction. This work was further stimulated by a conversation with a colleague, Dr Carol Jacobson, about her work with a small South American opossum and the role that the estrogen receptor (ESR) played in reproduction (Handa *et al.*, 1991). The serendipity of this conversation cannot be overlooked. Noting that this gene was closely linked on the human gene map to the human MHC it was decided to investigate further. Using a human ESR gene probe, available at the time through the American Tissue Type Culture catalogue, the first Southern restriction fragment length polymorphism (RFLP) analysis was performed on pig DNA and showed a clear segregation of two fragments, 4.3 and 3.7 kb (see Fig. 11.1, US Patent Number 5,374,526). These were called the A and B alleles. This allowed the gene to be mapped in the pig. Disappointingly for the hypothesis, the gene mapped to pig chromosome 1 not 7 (as had been predicted for linkage to MHC), but, on a hunch, the presence of these fragments was scored in 21 (of 22) Meishan females at ISU. The results were rather astounding as they showed a three-pig advantage to those with the A allele compared to those without it.

In an effort to obtain more funding and to pursue this interesting finding, it was decided to write a small industry grant. The grant was submitted to the National Pork Producer's Council with the preliminary results. Perhaps it was not too surprising that the significance of this

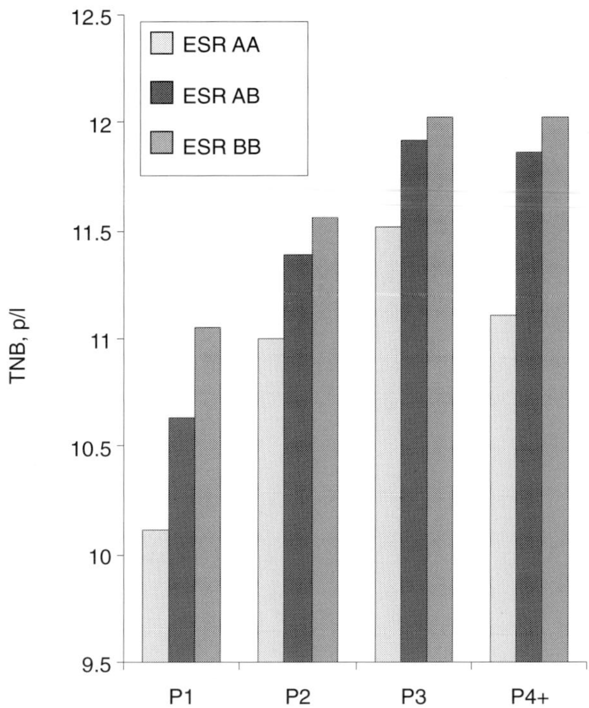

Fig. 11.1. The effect of ESR in LW-based purelines (total sows 4262), ESR BB sows had on average 0.47 more piglets than AB sows and 0.76 piglets more than AA sows.

small data set was questioned. The grant was refused with one reviewer, a prominent animal breeder stating, 'no single gene can control reproduction in the pig'. The stage was set that in order to continue the research more money was needed. The Biotechnology Research and Development Corporation (BRDC), a newly formed research management corporation that was a unique consortium that combined government, academia and the private sector, was approached next for funding. They were happy to consider the proposal if they could find interested companies that might be attracted to join the corporation and who could then exploit any intellectual property that might result. This of course would mean filing a patent in order to be able to license the technology and receive a return on investment. The Pig Improvement Company (PIC) showed real interest in the project. Their interest stemmed initially from their own use of Meishan germplasm that they had imported first into the UK and later into the USA. Further investigations with pure PIC Meishans using a limited number of full sisters revealed a difference of three pigs, similarly to the initial ISU results, and consequently excitement and interest in investigating the role of this gene in hybrid PIC lines grew. PIC joined BRDC and agreed to license the 'invention' once additional studies were completed. These studies produced early results that demonstrated that one ESR allele (called B), found first in the PIC Meishan pigs and later in the PIC Large White lines (Rothschild *et al.*, 1994), was associated with improved litter size in a Meishan × Duroc composite line. The results were similar to the first ESR results obtained at ISU but there the important allele was the alternate A allele (see Table 11.1). Based on what we now know, this difference was quite possibly real and is the result of different linkage between the polymorphism and a causative mutation (or quantitative brait loci, QTL – see also discussion below). However, it may simply be due to the small sample size in the ISU results with the apparent relationship being the result of chance. In the first stages of the research, there appeared to be an association within families, but not across the populations tested. These results, different associations between populations and between families, suggested a close linkage relationship between the marker and the causative mutation or QTL somewhere in the ESR region of pig chromosome 1. Nevertheless, the significant association of the ESR gene marker with litter size had real potential for commercial value and the patent application was pursued to protect the original invention of using ESR polymorphisms to improve litter size. It was also felt that a patent would provide additional support for these results in both the academic world and the marketplace. It should be noted that this work was initiated shortly after the identification and patenting of the method for testing and use of the 'Halothane' (HAL) gene. There was some debate in the scientific community as to the validity of the HAL patent, since the result seemed

Table 11.1. Summary of ISU/PIC results.

	Sow observations			Total number born			Number born alive			Parity	Reference
Population	AA	AB	BB	AA	AB	BB	AA	AB	BB		
ISU Meishan (MS)	7	*	14	12.7 ± 0.84	*	11.4 ± 0.71	12.4 ± 0.81	*	10.9 ± 0.65	1	US Patent
	5		11	14.2 ± 1.16		11.4 ± 1.31	12.8 ± 0.92		10.2 ± 1.17	2	5,374,526
	3		4	16.3 ± 0.33		13.5 ± 1.84	15.0 ± 1.53		13.3 ± 1.79	3	
PIC MS/ Duroc synthetic	32	41	12	10.9^{c}	12.4^{d}	12.2^{cd}	10.4^{g}	11.4^{h}	11.8^{h}	1	Rothschild *et al.*, 1994
PIC MS/Duroc synthetic	62	73	26	10.1^{a}	11.4^{b}	12.4^{b}	9.1^{a}	10.5^{b}	11.4^{b}	1	Rothschild *et al.*, 1996
PIC MS/Duroc synthetic	98	125	53	11.0^{a}	12.4^{b}	12.5^{b}	10.2^{a}	11.5^{b}	11.7^{b}	All parities	Rothschild *et al.*, 1996
PIC MS/Landrace synthetic	20	24	18	10.5^{c}	12.6^{d}	13.4^{d}	9.3^{c}	11.0^{cd}	11.9^{d}	1	Southwood *et al.*, 1995
PIC MS/Landrace synthetic	25	35	17	12.0	12.6	13.0	10.7	11.1	9.7	>1	Southwood *et al.*, 1995
PIC Large White and Large White synthetic lines	1023	2130	1109	10.1^{a}	10.6^{b}	11.0^{c}	9.4^{a}	9.9^{b}	10.2^{c}	1	Short *et al.*, 1997
PIC Large White and Large White synthetic lines	879	2330	1544	11.4^{a}	11.9^{b}	12.0^{b}	10.0^{a}	10.5^{b}	10.7^{b}		Short *et al.*, 1997
PIC Commercial sows	2678	1915	–	11.9^{a}	12.1^{b}	–				All parities	Unpublished

^{a,b,c}Different letters in the same row, $P<0.01$.
^{c,d}Different letters in the same row, $P<0.05$.
^{g,h}Different letters in the same row, $P<0.2$.
*AA and AB individuals were combined in this analysis.

obvious once the gene became a candidate (see Fujii *et al.*, 1991). Indeed, the HAL invention was predicted in publications from the MacLennan group where they set out the strategy for finding the mutation (MacLennan *et al.*, 1990). However, this opinion was based, at least in part, on a misunderstanding of the term 'obviousness' as required for patentability. Even so, it was possible at that time to find different patent agents willing to support or condemn the HAL patent. It should be noted that, subsequently, this patent was granted in several different territories (e.g. US 5,358,649, 25 October 1994). The ESR patent seemed less obvious and contrasted that of HAL, since despite the candidate gene approach no one had proven the role of genetic differences in the receptor and their relationship to litter size in the pig.

Later results (Rothschild *et al.*, 1994, 1996), again in Chinese pig crosses showed first parity differences between ESR homozygotes to be nearly 2.3 pigs per litter or an additive effect of an increase of 1.15 pigs per litter for each copy of the B allele (see Table 11.1). In subsequent litters, the effect of the B allele was about 0.5 pigs per litter and appeared to act in a dominant manner. In one of the data sets, an interesting interaction was observed with parity. Although the B allele was generally found to be associated with larger litters across parities, it was actually found to be associated with a drop in litter size in parity two (results not shown). That is, litter size for BB sows was lower in parity two than either AA or AB sows. This was initially puzzling, but results from nutritional studies using Meishan crosses provided insight into what was happening here, indicating that it was related in part to a general drop in litter size in the second parity at that time. The nutritional regime being used in the research herd had not been adjusted sufficiently to manage the increased litter size obtained from these 'exotic' Meishan crosses (see, for example, Sinclair *et al.*, 1999). The BB gilts producing the largest litters in parity one were most susceptible to this deficiency. Once this requirement was recognized, this effect was dealt with so that in subsequent analyses females with the B allele were consistently producing larger litters. This is an interesting example of genotype by environment interaction (G×E) and shows how it is possible to manage these differences in situations where there is sufficient understanding of the system. It is also interesting, as several people still consider that G×E is a reason not to utilize gene markers.

The B allele was also found to be segregating in several Large White populations (Rothschild *et al.*, 1995; Legault *et al.*, 1996; Rothschild *et al.*, 1996; Short *et al.*, 1997), in Chinese breeds (Chen *et al.*, 2000) and in a German synthetic line of Duroc and Large White origin (Drogemuller *et al.*, 2001), but the effect of the B allele seemed to differ somewhat between populations. However, in the largest study involving four commercial lines (Short *et al.*, 1997) differences for the B allele were about 0.42 pigs per litter or a difference of 0.84 pigs per litter between the AA and BB homozygotes. The effect was only slightly smaller in later parities where the B allele showed more of a dominant effect than an additive effect (see Fig. 11.1 and Table 11.1). In this study, no real line differences were seen but then all of the lines had many litter records. It is possible that the differences seen in some studies may be due to the relatively small numbers included (see below for an alternative explanation). This study also demonstrated that there were no significant pleiotropic effects on growth that might impact the commercial value of its use to select for larger litters. Smaller studies on ESR have shown some apparent differences for the association with litter size but these may be the result of sampling (e.g. AA animals had larger litters than AB animals in a Polish study (Korwin-Kossakowska *et al.*, 1999), and in F2 females from a Large White Meishan cross the highest litter size was seen for heterozygotes with BB gilts having significantly smaller litters than AA or AB gilts (van Rens *et al.*, 2001)). It is also possible, however, that these line differences may either indicate that the ESR locus is only a linked marker or the differences in genetic backgrounds between lines have an impact on the expression of the ESR gene. To date a causative mutation has not been described.

Having invested in the technology, PIC continued to investigate how best to utilize the ESR gene. One key question was whether the effect of the beneficial B allele would be expressed in commercial crossbred sows. In order to test the hypothesis that ESR would explain variation in litter size in this situation, samples for marker testing were taken from commercial farms with reproductive performance recording. The sows used were produced from a cross, including one parent from the lines that showed an effect as presented by Short *et al.* (1997). A total of 4493 sows producing 14,827 litters were sampled from nine farms. The frequency of the B allele was relatively low as the sow mothers were from a line essentially fixed for the A allele. Overall there was a difference of approximately 0.2 piglets between the litters of AA and AB sows ($P<0.01$) with the heterozygous sows having the larger litters as predicted from the pure line results, although the size of the effect is about half that reported for the pure lines (Table 11.1, Short, McLaren, Wilson and Plastow, unpublished results). Interestingly, there was a significant interaction with farm, with the size of the ESR effect tending to be larger the better the reproductive performance of the farm, and in one case the effect was in the opposite direction (i.e. the AB sows had smaller litters). One might speculate that this observation is related to the nutrition effect initially observed in Meishan crosses (see above), indeed, when analysed by parity the ESR effect was reduced in parity to two sows. These differences may explain the lower estimate for the size of the effect, but alternative explanations are also possible including epistatic effects. The smaller effect observed in this large data set was then used by PIC to estimate the value of the technology for the producer using cross-bred sows selected for ESR. An estimate of the value of the ESR marker test can be derived by considering the impact of fixing the preferred allele in a population. Here, we consider the impact in the USA for customers of PIC. The calculation needs to take into account the starting frequency of the allele, as well as the size of the effect. As we have shown above, the effect in commercial sows derived from PIC Large White pigs was on average around 0.2 pigs per litter per copy of the preferred B allele. The initial allele frequency was 0.5, so that if we assume that the allele is only derived from the Large White component of the cross (a reasonable assumption as the ESR marker is not segregating in Landrace lines that are used in the cross-bred sow), then the frequency of AB parent sows (commercial sows) would also be 0.5. If we assume that each sow will have five litters per lifetime and an additional pig per litter raised generates around \$30 additional margin for the producer, then fixing ESR will generate an extra \$30 per sow (0.2 pigs × 5 litters × \$30). In 1996, PIC sold about 700,000 parent gilts in the USA, thus by fixing the gene in PIC Large White lines this would, given the gene frequency of 0.5, have been worth an extra \$10.5 million to the US industry (0.5 × \$30 × 700,000). Additional value would of course come from worldwide sales.

A Description of the Invention

The ESR patents are for genetic markers for litter size, methods for identifying such markers and methods for screening pigs to determine those more likely to produce larger litters and preferably selecting those pigs for future breeding purposes based on the ESR polymorphisms. The markers are based on the presence or absence of certain polymorphisms in the pig ESR gene. Preferably, the polymorphism is an RFLP polymorphism. The original application (Ser. No 07/687,708) was filed 19 April 1991, and was abandoned due to significant unresolved disagreements with the patent examiner and a failure of the original lawyers to describe the invention properly. A general observation is that patent examiners are somewhat lacking in general knowledge with respect to animal breeding and quantitative genetics. This was to be expected, as patents had not generally been filed in this field (see also Introduction), however, it presented some interesting challenges during the patent process for this gene. These problems were addressed by revision and refiling of the application. A continuation-in-part was filed on 19 October 1991 (Appl. No. 961,819). The first patent to be granted is US Patent No. 5,374,526 issued on 20 December 1994 and this was followed by US Patent No. 5,550,024 issued 27 August 1996 which was a continuation in part (C-I-P) of the earlier granted patent. This C-I-P was filed with a specific example of the use of polymerase chain reaction

(PCR), which provided a more general approach and provided the basis for broader claims. In part, this was required, as the claims allowed by the examiner of US 5,374,526 were somewhat restricted. It should be noted that the filing of continuations is a specific difference between the US Patent Office and patent offices in other countries. It was possible to file a C-I-P to include the specific sequence of the polymorphism together with primers for PCR amplification of the region in order to broaden the coverage of the case. This would not have been possible, for example, in Europe and it is interesting to note that when the European application was eventually granted (EP No. 0580767, December 2000) the claims allowed were relatively broad and essentially as originally filed in the USA. Additional patents have been applied for and have been granted in various locations worldwide.

The original inventors were Max Rothschild and Carol Jacobson (see above), and David Vaske and Christopher Tuggle, as well as Alan Mileham and Graham Plastow at PIC for the C-I-P, where they contributed significantly with the creation of the marker test. Rights to market the technology are jointly owned by BRDC and ISU, who granted an exclusive licence to PIC for worldwide use of the technology. It should be noted that PIC strongly supported the project, were directly involved in the research, provided indirect funding for the research and finally participated in completing the invention for the US patent (see above).

The breadth of the coverage provided by the patent can probably only be determined in practise. At the time the patent was filed we considered that the invention related to the identification of the association between variation in the ESR region of the genome and variation in litter size, thereby providing the basis for using genetic markers to select for more prolific pigs. This can be interpreted to include other markers in the region (linked markers) that can be used for this purpose (as claimed). However, an analysis of this aspect is beyond the scope of this chapter and, in any case, we would anticipate that the nature of what is allowed will evolve over time as the field develops such that any view would not necessarily be generally applicable.

Research Implications

Implicit in the granting/publishing of a patent is that it will encourage future developments in the same field. Certainly this has been the case with the ESR patents. Since the discovery of the association of the ESR genotype with litter size, different groups have investigated this gene (see section describing ESR) and several other individual genes have been chosen for candidate gene studies for reproductive traits. In addition, there was an upsurge in the use of the candidate gene approach in pigs, although many groups have preferred to focus on the use of QTL analyses based on divergent crosses (see Rothschild and Soller, 1997, for a discussion of the different approaches and in particular a defence of the candidate gene approach). Of course, as the amount of genome information increases, it is likely that the most efficient approach may be a combination of the two approaches (see Ciobanu *et al.*, 2001). Markers associated with growth and appetite, meat quality, muscling and coat colour have been identified and patents applied for (Kim *et al.*, 2000; Milan *et al.*, 2000; Nezer *et al.*, 1999; Marklund *et al.*, 1998).

Examples of candidate gene work for litter size (Rothschild *et al.*, 1997) include retinol binding protein 4 (RBP4; Rothschild *et al.*, 2000), prolactin receptor (PRLR; Vincent *et al.*, 1998), and the gene encoding the beta subunit of follicle stimulating hormone (FSHB; Li *et al.*, 1998). These genes were selected due to their role in reproductive trait physiology. Additional patents have been filed and granted to cover these genes affecting litter size in the pig. These include prolactin receptor (PRLR), retinol binding protein 4 (RBP4) and FSHB under US Patent Nos 5,935,784, 5,939,264 and 6,291,174, respectively.

Many studies have produced positive results with RBP4 (Messer *et al.*, 1996; Rothschild *et al.*, 2000), PRLR (Vincent *et al.*, 1998; Drogemuller *et al.*, 2000; Isler *et al.*, 2000) and FSHB (Li *et al.*, 1998; Huang *et al.*, 2000). Other research has had negative results (Linville *et al.*, 2001). However, in this study the number of pigs analysed for each genotype ranged from 189 (EGF) to 524 (PRLR) individuals, with litter size records taken only on 244 first-parity gilts. This sample population was likely insufficient in size to estimate accurately effects of the candidate genes. Short *et al.* (1997) indicated that over 1000 litter records were necessary to obtain a consistent estimate for the effect of ESR. So interpretation of results from small-scale association studies must always be done with care.

Early genome scans (Wilkie *et al.*, 1996; Rohrer *et al.*, 1999; Cassady *et al.*, 2001) for reproductive traits have also been employed using crosses between Meishan and Large White breeds several QTL were found. These genome scans have been useful to identify regions of interest but they have not had the statistical power to find moderate to small effects. Furthermore, they have not had statistical power equal to most of the candidate gene analyses and therefore do not by themselves refute findings from larger candidate gene analyses involving several thousands of litter records. Interestingly, none of the QTL detected corresponded to the candidate genes discussed above. Finally, if ESR was segregating in these populations and at an intermediate allele frequency then there would not be the ability to find ESR gene effects in scans where ESR was not included in the analysis. It is unfortunate that researchers conducting these analyses have in general failed to understand the power of their experiments and other limitations that might have prevented them from seeing these candidate gene effects.

Responses from the Public and the Industry

In 1994, the ESR story was first told to a very large audience of interested researchers at the 5th World Congress on Genetics Applied to Livestock Production in Guelph, Canada. The first question, from a molecular geneticist was 'Why did you patent the marker test?', although, as has been noted, the first pig DNA marker was identified and patented by Canadian researchers in 1990 (Fujii *et al.*, 1991). Other questions reflected a real interest in the technology but some suggested doubt in the results and even disgust that industry was involved and a patent application had been filed. Clearly the most surprising and perhaps disappointing aspects of the ESR patent story for the inventors were some of the generally negative remarks received from some individuals. Although the exploitation of the invention has been a success (see below), these negative comments tended to get the most attention. Comments fell into several different groups, for example from well-meaning and or naïve researchers concerned about patenting to researchers and industry personnel unhappy over the exclusivity of the licensing arrangement. Some of these responses are discussed below.

Supporters of the technology were intrigued by the candidate gene approach but had concerns about the statistical power of such an approach. Many researchers wondered if this was just a linkage effect. To date this and the fact that a causative mutation has not been made public continues to be a criticism of the patent. This misses the point that the invention is the identification of an association between polymorphisms in ESR and litter size so as to enable the selection of pigs with the potential for larger litters. The nature of the described polymorphism is irrelevant in this respect, although it is of genuine scientific interest and several groups have searched for additional polymorphisms within the gene (e.g. Drogemuller *et al.*, 1997).

The notoriety that the patent attracted caused many traditional animal breeders to take notice of patents and molecular biology. One group of respected university people expressed their concerns that academia should not be involved in a moneymaking enterprise. Furthermore, they felt that private appropriation of knowledge that had been generated in public institutions would erode the free flow of ideas. These can certainly be viewed as legitimate concerns (and are considered so by the authors) but they ignore the fact that university researchers get less money from public sources than they did several years ago, and the requirement to get more research dollars has changed how they operate. Farm animal research was catching up with the rest of biology in this respect. One of the intended aims of the patent system is specifically to stimulate the flow of ideas (or at least to provide access to new ideas) in a way that rewards inventive effort in terms of some form of commercial success.

Other individuals had complaints that DNA or animals could not be patented, and they questioned the exclusive nature of the patent/licensing agreement. The patenting of animals is not permitted in most of the major jurisdictions and this has not been attempted with the ESR patent. In livestock, other gene patents existed in 1994 but none had involved discoveries involving exotic germplasm. One Chinese researcher complained that the US researchers were 'exploiting' China by taking natural resources and patenting them. Whilst the Meishan breed was useful in the identification of the polymorphism, the breed itself was not subject to the patent claims. Clearly one of the misunderstandings was that the gene had been patented, whereas in this case a 'methods patent' had been granted that described the use of the gene, which is not restricted to the Meishan breed. This was a general source of misunderstanding for several years. Several discussions were held at industry meetings in Europe where opponents thought that the patent stopped them using their pigs if they carried the B allele of ESR so that they were determined to try to prevent patents on gene markers. This misunderstanding was eventually corrected and agreement on the value of such patents was reached among industry representatives.

Another well-respected animal breeder, Daniel Gianola, initiated an e-mail campaign (this story is provided with permission and encouragement by Dr Gianola) involving a character called Poul Henningsen, a Danish pig farmer that, writing in broken English, wanted to know if Prof. Roskilde could give him a discount for litter size genes. This e-mail spoof was to draw attention to concerns about the move from initial public release of results to patenting and did generate some useful exchange of ideas. Furthermore, Dr Gianola commented on other standard activities that animal breeders do including calculation of a mean and wondered if some scientist could have patented it. This argument is expanded in the accompanying review in this book by Schaeffer concerning the patenting of the test day model.

Perhaps the most disappointing aspect has been the limited research undertaken to explore or dispute the role of ESR in the reproductive processes. Even so, there have been some attempts (Isler *et al.*, 1999a,b; Vonnahme *et al.*, 1999; van Rens, 2001; van Rens *et al.*, 2001). One of the problems is the high cost and difficulty in undertaking this type of study, which in turn means that the size of such studies is very likely to be limited and experience suggests that relatively large numbers are required for studies of litter size. Vonnahme *et al.* (1999) reported that they could not find any evidence to support a role for ESR in controlling litter size, but small experiments of this type do not have the statistical power to declare 'no effect' of a gene. A larger research effort, although still restricted, was carried out by van Rens and van der Lende whose experiments looking at ESR (and also PRLR) investigated where in pregnancy ESR might exert its effect (van Rens, 2001). They concluded that the genes (or linked QTL) influenced different components of litter size. For ESR this was an effect on fetal survival (apparently from an effect on placental size); however, it should be noted that in one part of the study the A allele appeared to be associated with larger litters (van Rens *et al.*, 2001). In the case of PRLR, the effect was on ovulation rate. Interestingly, Isler *et al.* (2000) also concluded that PRLR has a significant effect on number of embryos. This group also looked at the role of ESR and its relationship to the components of litter size, concluding that BB animals had 0.89 more fetuses per horn at day 75 of gestation than AA females ($P<0.06$) and other traits, including fetal survival, total uterine length, total fetal weight and ovulation rate within horn, also displayed favourable, but not statistically significant trends with respect to ESR BB genotype (Isler *et al.*, 1999a,b).

Those in the swine industry approached the discovery and patent with different viewpoints depending on whether they felt entitled to reap in the benefits or were threatened economically by the discovery. The Iowa Pork Producers Association fought a protracted battle with ISU claiming that their money, which helped allow importation of the Meishan pigs to ISU, gave them rights to a share of future income from the discovery. This disagreement was finally settled after they were shown that the discovery was independent of the pigs. Many swine genetics companies were understandably unhappy with the exclusivity of the licence agreement. One company told customers that 'it doesn't work – we tried it', which suggested that if they really had tested the genetic markers that they were infringing the patent. Another company's swine breeder noted that ESR was linked to the major gene for litter size and because of its exclusive licensing arrangement suggested attempting to avoid infringement by stating 'fortunately for others, there are several potentially alternative markers in the same region of the chromosome that are publicly available' (Webb, 1999). As discussed earlier in the chapter, this point is open to interpretation, but the view expressed by Webb is counter to that of several experts in the field.

Conclusions

In the writing of this chapter, a number of thoughts surfaced relative to the ESR story that had been forgotten by the authors. Certainly the pressure to secure funding from BRDC and to get support from PIC made patenting a reasonable alternative to the usual university approach of 'write a grant that may or not be funded and then publish the work and see if anyone uses it'. At the time, a large factor in pursuing the patent was that it represented a first for Rothschild and his co-inventors. It therefore had the positive feel of a new venture and provided new possibilities in the future for application of their work.

Another overriding fact was the desire to make clear that the discovery was unique and certainly a different result for animal breeding, in that a single gene had a large influence on a quantitative trait. Even today, ESR continues to be an interesting story from a scientific, industrial and personal basis. The commercialization of ESR led to it being awarded an R&D100 award in 1999, presented by the R&D magazine (September 1999, Vol. 41) which reviews inventions and their commercialization in the USA. This was a rare honour as most of the inventions recognized are not of a biological nature but more often engineering related. ESR continues to be marketed and customers of PIC from around the world are choosing both females and males with known ESR genotypes. Its success continues to grow and while there are detractors it still is one of the very few genetic markers associated with a quantitative trait. It has led to other genetic markers for litter size being discovered and most importantly to some rethinking of the notion that favourable alleles from single genes with large effects would be expected to be fixed in the population. Hopefully, this chapter gives a flavour for some of the difficulties and problems surrounding the patenting of this type of invention. Overall, despite these difficulties, the process has been a positive learning and rewarding experience.

Acknowledgements

This work was financially supported by BRDC, PIC Group and the Iowa Agriculture and Home Economics Experimental Station, Ames, paper no. J-19687, project no. 3600, as well as by Hatch Act and State of Iowa funds.

The authors wish to thank their co-inventors for development of the ESR discovery, and intellectual property staff of ISURF at Iowa State University and Ms Heidi Nebel, the attorney of record, for guidance and help in securing the ESR patents along with the UK patent agent Paul Chapman, who also helped with comments on our discussion of the patent process. The early enthusiasm of David McLaren was an important contribution and his help, together with Tom Short in planning and acquisition of animal samples and information, is also appreciated. Suggestions and remarks on the manuscript by Chris Haley are greatly appreciated.

References

Cassady, J.P., Johnson, R.K., Pomp, D., Rohrer, G.A., Van Vleck, L.D., Spiegel, E.K. and Gilson, K.M. (2001) Identification of quantitative trait loci affect reproduction in pigs. *Journal of Animal Science* 79, 623–633.

Chen, K., Huang, L.S., Li, N. and Luo, M. (2000) The combined genotypes effect of *ESR* and *FSHB* genes on litter size traits in five different pig breeds. *Chinese Science Bulletin* 145, 1963–1966.

Ciobanu, D.C., Bastiaansen, J., Malek, M., Helm, J., Woollard, J., Plastow, G.S. and Rothschild, M.F. (2001) Evidence for new alleles in the protein kinase AMP-activated, subunit gene associated with low glycogen content in pig skeletal muscle and improved meat quality. *Genetics* 159, 1151–1162.

Conley, A.J., Jung, Y.C., Schwartz, N.K., Warner, C.M., Rothschild, M.F. and Ford, S.P. (1988) Influence of SLA haplotype on ovulation rate and litter size in miniature pigs. *Journal of Reproduction and Fertility* 22, 595–601.

Drogemuller, C., Thieven, U. and Harlizius, B. (1997) An AvaI and a MspA1I polymorphism at the porcine oestrogen receptor (ESR) gene. *Animal Genetics* 28, 58–71.

Drogemuller, C., Hamann, H., Krieter, J., Preshun, U., Wallenburg, J. and Distl, O. (2000) Candidate gene markers connected to litter size in German pigs. *International Society of Animal Genetics*, Abstract B016.

Drogemuller, C., Hamann, H. and Distl, O. (2001) Candidate gene markers for litter size in different German pig lines. *Journal of Animal Science* 79, 2565–2570.

Fujii, J., Otsu, K., Zorzato, F., de Leon, S., Khanna, V.K., Weiler, J.E., O'Brien, P.J. and MacLennan, D.H. (1991) Identification of a mutation in porcine ryanodine receptor associated with malignant hyperthermia. *Science* 253, 448–451.

Handa, R.J., Rodriguez, E.W., Fox, C.A. and Jacobson, C.D. (1991) Characterization and distribution of estrogen receptors in the diencephalon of the gray short-tailed opossum. *Brain Research* 539, 6–10.

Huang, L.S., Chen, K.F., Li, N., Ren, J., Ding, N.S., Meyer, J.N., Besenko, S.P., Gao, J. and Luo, M. (2000) Genetic variation at the type I and type II markers loci in different Chinese local and Western commercial pig breeds. *International Society of Animal Genetics*, Abstract C031.

Huang, M.Y. and Rasmusen, B.A. (1982) Parental transferrin types and litter size in pigs. *Journal of Animal Science* 54, 757–762.

Imlah, P. (1970) Evidence for the Tf locus being associated with an early lethal factor in a strain of pig. *Animal Blood Groups and Biochemical Genetics* 1, 5–13.

Isler, B.J., Irvin, K.M., Neal, S.M., Moeller, S.J. and Davis, M.E. (1999a) Association between the estrogen receptor gene and reproductive components in swine. *Journal of Animal Science* 77(Suppl. 1), 32.

Isler, B.J., Irvin, K.M., Neal, S.M., Moeller, S.J., Davis, M.E. and Meeker, D.E. (1999b) The effect of estrogen receptor genotype, breed, and parity on litter traits and reproductive tract traits in swine. *Journal of Animal Science* 77(Suppl. 1), 131.

Isler, B., Irvin, K.M., Rothschild, M.F. and Evans, G.J. (2000) Association between the prolactin receptor gene and reproductive traits in swine. *International Society of Animal Genetics*, Abstract C032.

Jensen, E.L., Smith, C., Baker, L.N. and Cox, D.F. (1968) Quantitative studies on blood group and serum protein systems in pigs. II. – Effects on production and reproduction. *Journal of Animal Science* 27, 856–862.

Kennedy, B.W., Moxley, J.E. and Saison, R. (1973) Some relationships between blood group factors and economic traits in swine. *Canadian Journal of Animal Science* 53, 389–397.

Kim, K.S., Larsen, N., Short, T., Plastow, G. and Rothschild, M.F. (2000) A missense variant of the porcine melanocortin-4 receptor (MC4R) gene is associated with fatness, growth, and feed intake traits. *Mammalian Genome* 11, 131–135.

Korwin-Kossakowska, A., Kuryl, J. and Pierzchala, M. (1999) An analysis of relations between the polymorphism of estrogen receptor gene and some reproduction traits in Zlotnicka Spotted × Polish Large White pigs. *Animal Science Papers and Reports* 17, 155–161.

Kristjansson, F.K. (1964) Transferrin types and reproductive performance in the pig. *Journal of Reproduction and Fertility* 8, 311–317.

Legault, C., Gruand, J., Lebost, J., Garreau, H., Ollivier, L., Messer, L.A. and Rothschild, M.F. (1996) Frequence et effect sur la prolificite du gene ESR dams deux lignees Large White en France. *Journées Recherches Porcine en France* 28, 9–14.

Li, N., Zhoa, Y.F., Xiao, L., Zang, F.J., Chen, Y.Z., Dai, R.J., Zang, J.S., Shen, S.Q., Chen, Y.F. and Wu, C.X. (1998) Candidate gene analysis for identification of genetic loci controlling litter size in swine. *Proceedings of the 6th World Congress on Genetics Applied Livestock Production* 26, University of New England, Armidale, New South Wales, pp. 183–186.

Linville, R.C., Pomp, D., Johnson, R.K. and Rothschild, M.F. (2001) Candidate gene analysis for loci affecting litter size and ovulation rate in swine. *Journal of Animal Science* 79, 60–67.

Marklund, S., Kijas, J., Rodriguez-Martinez, H., Rönnstrand, L., Funa, K., Moller, M., Lange, D., Edfors-Lilja, I. and Andersson, L. (1998) Molecular basis for the dominant white phenotype in the domestic pig. *Genome Research* 8, 826–833.

MacLennan, D.H., Duff, C., Zorzato, F., Fujii, J., Phillips, M., Korneluk, R.G., Frodis, W., Britt, A. and Worton, R.G. (1990) Ryanodine receptor gene is a candidate for predisposition to malignant hyperthermia. *Nature* 343, 559–561.

Messer, L.A., Wang, L., Legault, C. and Rothschild, M.F. (1996) Mapping and investigation of candidate genes for litter size in French Large White pigs. *Proceedings of the 25th International Conference on Animal Genetics*, Tours, France.

Milan, D., Jeon, J.T., Looft, C., Amarger, V., Robic, A., Thelander, M., Rogel-Gaillard, C., Paul, S., Iannucelli, N., Rask, L., Ronne, H., Lundstrom, K., Reinsch, N., Gellin, J., Kalm, E., Roy, P.L., Chardon, P. and Andersson, L. (2000) A mutation in *PRKAG3* associated with excess glycogen content in pig skeletal muscle. *Science* 288, 1248–1251.

Nezer, C., Moreau, I., Karim, L., Brouwers, B., Coppieters, W., Detilleux, J., Hanset, R., Kvasz, A., Leroy, P. and Georges, M. (1999). An imprinted QTL with major effects on muscle mass and fat deposition maps to the IGF2 locus in pigs. *Nature Genetics* 21, 155–156.

O'Brien, S.J. (1993) *Genetic Maps*, 6th edition, Book 4. *Nonhuman Vertebrates.* Cold Spring Harbor Laboratory Press, Cold Spring Harbor, New York.

Rasmusen, B.A. and Hagen, K.L. (1973) The H blood-group system and reproduction in pigs. *Journal of Animal Science* 37, 568–573.

Rohrer, G.A., Ford, J.J., Wise, T.H., Vallet, J.L. and Christenson, R.K. (1999) Identification of quantitative trait loci affecting female reproductive traits in a multigeneration Meishan–White composite swine population. *Journal of Animal Science* 77, 1385–1391.

Rothschild, M.F. (1996) Genetics and reproduction in the pig. *Animal Reproduction Science* 42, 143–151.

Rothschild, M.F. and Soller, M. (1997) Candidate gene analysis to detect traits of economic importance in domestic livestock. *Probe* 813–820.

Rothschild, M.F., Zimmerman, D.R., Johnson, R.K., Venier, L. and Warner, C.M. (1984) SLA haplotype differences in lines of pigs which differ in ovulation rate. *Animal Blood Groups and Biochemical Genetics* 15, 155–158.

Rothschild, M.F., Jacobson, C., Vaske, D.A., Tuggle, C.K., Short, T., Sasaki, S., Eckardt, G.R. and McLaren, D.G. (1994) A major gene for litter size in pigs. *Proceeding of the 5th World Congress on Genetics Applied to Livestock Production* 21. University of Guelph, Guelph, Ontario, Canada, pp. 225–228.

Rothschild, M.F., Vaske, D.A., Tuggle, C.K., Messer, L.A., McLaren, D.G., Short, T.H., Eckardt, G.R., Mileham, A.J. and Plastow, G.S. (1995) Estrogen receptor locus is a major gene for litter size in the pig. *Proceedings of the European Association of Animal Production*, Wageningen, The Netherlands, p. 53.

Rothschild, M.F., Jacobson, C., Vaske, D.A., Tuggle, C.K., Wang, L., Short, T., Eckardt, G.R., Sasaki, S., Vincent, A., McLaren, D.G., Southwood, O., van der Steen, H., Mileham, A. and Plastow, G. (1996) The estrogen receptor locus is associated with a major gene influencing litter size in pigs. *Proceedings of the National Academy of Sciences USA* 93, 201–205.

Rothschild, M.F., Messer, L.A. and Vincent, A. (1997) Molecular approaches to improved pig fertility. *Journal of Reproduction and Fertility* 52 (Suppl.), 227–236.

Rothschild, M.F., Messer, L.A., Day, A., Wahs, R., Short, T., Southwood, O. and Plastow, G. (2000) Investigation of the retinol binding protein (RBP4) gene as a candidate gene for litter size in the pig. *Mammalian Genome* 11, 75–77.

Short, T.H., Rothschild, M.F., Southwood, O.I., McLaren, de Vries, A., van der Steen, H., Eckardt, G.R., Tuggle, C.K., Helm, J., Vaske, D.A., Mileham, A.J. and Plastow, G.S. (1997) Effect of the estrogen receptor locus on reproduction and production traits in four commercial pig lines. *Journal of Animal Science* 75, 3138–3142.

Sinclair, A.G., Cia, M.C., Hoste, S. and Edwards, S.A. (1999) Response to dietary protein during lactation of Meishan synthetic and European White sows fed to attain two levels of backfat level at farrowing. *Animal Science* 69, 583–590.

Southwood, O.I., van der Steen, H.A.M., Mileham, A.J., Plastow, G.S. and Cuthbert-Heavens, D. (1995) Evaluation of the oestrogen receptor (ESR) gene in Meishan synthetic pigs. *Proceedings of the 46th Annual Meeting European Association of Animal Production* 1, 53 abstr.

Vaiman, M., Renard, C.H. and Bourgeux, N. (1988) SLA, the major histocompatibility complex in swine: its influence on physiological and pathological traits. In: Warner, C., Rothschild, M. and Lamont, S. (eds) *The Molecular Biology of the Major Histocompatibility Complex of Domestic Animal Species*. Iowa State Press, Ames, Iowa, pp. 23–38.

van Rens, B.T.T.M. (2001) Physiological aspects of two candidate genes for litter size in pigs. PhD thesis, Department of Animal Science, Wageningen University, The Netherlands.

van Rens, B.T.T.M., de Groot, P.N. and van der Lende, T. (2001) The effect of the estrogen receptor genotype on litter size and placental traits at term in F2 crossbred gilts. *Theriogenology*, in press.

Vincent, A.L, Evans, T.H., Short, T.H., Southwood, O.I., Plastow, G.S., Tuggle, C.K. and Rothschild, M.F. (1998) The prolactin receptor gene is associated with increased litter size in pigs. *Proceedings of the 6th World Congress Genetics Applied Livestock Production* 27, University of New England, Armidale, 15–18.

Vonnahme, K.A., Wilson, M.E. and Ford, S. (1999) Estrogen receptor is not associated with placental size in the pig. *Journal of Animal Science* 77(Suppl. 1), 71.

Warner, C.M. and Rothschild, M.F. (1991) The swine major histocompatibility complex (SLA). In: Srivasava R. (ed.) *Immunogenetics of the MHC*. Verlag Chemie International Publications, New York, pp. 368–397.

Webb, J. (1998) *Journal of Dairy Science* 81(Suppl. 2).

Webb, J. (1999) Objectives and strategies in pig improvement: an applied perspective. *Journal of Dairy Science* 81, 36–46.

Wilkie, P.J., Paszek, A.A., Flickinger, G.H., Rohrer, G.A., Alexander, L.J., Beattie, C.W. and Shook, L.B. (1996) Scan of 8 porcine chromosomes for growth, carcass and reproductive traits: two likely quantitative trait loci. *Proceedings of the 25th International Conference on Animal Genetics*, Abstract E068. Blackwells, Oxford, UK.

12

Development and Commercialization of a Genetic Marker for Marbling of Beef in Cattle: a Case Study

William Barendse

Molecular Animal Genetics Centre, CSIRO Livestock Industries, Level 3 Gehrmann Laboratories, Research Road, University of Queensland, St Lucia, QLD 4072, Australia

Introduction

DNA diagnostic tests have begun to alter significantly the practice of genetic prediction, for example in human genetic counselling for congenital disease, for risk for cancer and for the practice of animal breeding. Although the Guthrie style tests for some congenital disorders have been available for decades, as well as protein tests for sickle cell anaemia and a handful of other disorders, the modern period of readily available tests started with the identification of the mutation causing cystic fibrosis. This has been followed by a large number of DNA tests for congenital disorders as well as genetic susceptibilities for cancer (Collins, 1995). There has been a time lag in availability of such tests for animals. There are DNA tests for specific congenital disorders (e.g. BLAD, Pompe's disease, double muscling; OMIA, 2002) but up to now there have been few fully commercialized DNA tests for a 'susceptibility' or where variation at many genes and the environment affects the characteristic, i.e. tests for genes of moderate effect called quantitative trait loci (QTL).

One of the aims of molecular genetics in cattle is to identify the genetic variation that contributes to economically important traits of livestock so that breeding decisions can be based on an understanding of the ultimate genetic material, DNA. For tests that detect congenital mutations, such as double muscling in cattle, the testing and interpretation is simple. You either have or have not got one or more of the mutations in the myostatin gene which determines whether the animal has the double muscling condition (e.g. Grobet *et al.*, 1997). Something as simple as that makes for simplicity also in commercialization, since a deep understanding of genetics is not usually required of the commercializer and the tests result in an easy categorization of individuals. The situation is more complex in traits affected by many genes in which genetic variation specifies, say, a susceptibility or contributes to a genetic potential, where the risks are expressed in probabilities or likelihoods and, for diseases, a threshold may need to be passed before symptoms are seen. This implies a sophisticated understanding of population genetics by the commercializer who will need to integrate information on several genotypes as well as information on the environment.

Most of the genetic tests that will be developed in the future will be for susceptibilities or genetic potentials, not only in humans but also in other animals. In humans, most of the congenital mutations for which DNA tests have still to be developed are extremely rare or sporadic. In livestock, due to the nature of selection by farmers, lines that carry congenital mutations are penalized, and mild inbreeding and the removal of runts, performed over thousands of years, have to some extent scrubbed congenital mutations from livestock or bloodstock germlines. Rather, it is the genes that contribute to genetic potential for production quality that are still largely untapped and they represent not only a frontier for science but also a frontier for commercialization, since commercializers are in general inexperienced in this particular technology.

Description of the Invention Including Its Uses

This invention is a series of DNA tests that can be applied to DNA from any tissue source of the animal and from which a trained practitioner can determine: (i) a set of genotypes and (ii) from those genotypes can predict the marbling potential of an animal or the average marbling performance of groups of animals with different genotypes (Barendse, 1997). Most current DNA tests of organisms will provide a genotype which can be related directly to the observed characteristic of an animal, and the possession of the genotype gives the observed characteristic: there is little grey area. Most of these tests are for genetic variants that have large effects on the animal's observed body or phenotype and in general are deleterious mutations (Collins, 1995; see lists in OMIM, 2002). However, the series of DNA tests described here are for genetic variants that have much smaller individual effects; they are usually subsumed under the title of individual genetic variation, they have complex aetiology and represent most of the genetic variation that one considers to be normal. Finding the genes for such normal variation is difficult and the difficulties are only now being overcome in a systematic way. These difficulties also exist in patenting and commercialization of the information, as will be seen below. Nevertheless, increasingly, other tests like these will be commercialized to address some of the most important medical and agricultural characteristics of animals (e.g. Todd *et al.*, 1987; Lynch and Watson, 1992; Coleman *et al.*, 1995; Lembertas *et al.*, 1997).

This invention deals with marbling in beef, so it is appropriate to describe the trait to some extent, with particular emphasis on why a molecular genetic approach was required. Marbling is visible intramuscular fat which contributes to tenderness and flavour of beef and underpins quality grades in several countries (e.g. AUS-MEAT, 1998; MSA, 2002). Higher marbling scores are associated not only with increased consumer satisfaction but also with increased profit to the beef industry. The degree of marbling is affected not only by the environment but it is also expected to be influenced by many genes, some of which may have a significant individual effect on marbling. Only two genetic effects appear to have a 'congenital' effect on marbling, both serving to reduce it drastically, which is not wanted. These are mutations at the myostatin gene associated with double muscling and the presence or absence of a Y chromosome (Hanset *et al.*, 1982; Worrell *et al.*, 1987). No congenital mutations have yet been described that radically increase marbling score in cattle. Higher marbling scores result from feeding cattle excess energy in feedlots and the differences in individual performance are usually attributed to individual and breed genetic differences.

Marbling score is difficult to measure or to improve genetically which provides a rationale for developing DNA diagnostic tests. Firstly, marbling score is a categorical trait (Sokal and Rohlf, 1981) scored to a visual standard by a trained assessor, so there is a subjective element to it as well (AUS-MEAT, 1998). Secondly, it is usually measured *post mortem*, in the chilled carcass, which makes genetic improvement difficult. Thirdly, bulls show only trace amounts (Savell *et al.*, 1986) of marbling which means that expensive progeny tests are required as surrogates for measuring the animal itself. Proxies for marbling score can be obtained from ultrasound images of scanned muscle (Brethour, 2000) but the correlation to chiller estimates leaves room for significant misclassification. Marbling score represents an underlying level of intramuscular fat, but measurement of chemically extracted intramuscular fat is even more time consuming and expensive than progeny tests, although there is less of the subjective element to them. Given the difficulty of assessing an animal's marbling characteristics on the one hand and the high value of having such information on the other, we felt that DNA markers for marbling were an ideal candidate for commercialization. If a test for marbling could not be commercialized, it would be difficult to see how DNA markers for less desirable attributes of animals would be commercialized.

We approached the identification of genes for marbling using standard systematic screening of the bovine genome in conjunction with a few heuristic devices to narrow down the regions in which the search would occur. One method of tracking down unidentified genes is first to locate them to the gene map and then progressively refine their location until a DNA sequence can be obtained (Woolf, 1955; Thoday, 1961; Soller *et al.*, 1976; Todd *et al.*, 1987; Paterson *et al.*, 1988). The other method is the conventional biochemical approach to purify the gene product and then perform experiments to characterize the gene function. Since the action of 'marbling genes' (i.e. a gene one of whose functions affects marbling score) was known but no candidate gene products were available, only the first option was truly viable. Thus individual differences in marbling scores were used to locate the genes for marbling to chromosomes. However, this is a monumental task at the best of times so some heuristic methods were used primarily to shorten the task and reduce the cost to a reasonable budget. The pressures of time and cost arose because there was international competition to identify these genes or, at least, predictive DNA markers. As part of the genetic resources for the project, we were able to get commercial cattle of the appropriate specifications including a single family of 594 offspring of a single Wagyu bull that were going through the feedlot as a group. Heuristic methods, or rules of thumb, in gene mapping are ways in which one can avoid a systematic search that essentially begins at the centromere of chromosome 1 and works methodically through until one reaches the telomere of the Y chromosome. This method is slow but sure. If one has no idea of the genes, one could focus attention on the gene-rich regions first, since a large part of the genetic content of mammalian genomes is contained in relatively few regions (Antonarakis, 1994; IHGSC, 2001). Nevertheless, I chose a somewhat different heuristic. Dekkers and Dentine (1991) had shown that a series of highly polymorphic DNA markers, one per chromosome, placed as close to the middle of a linkage group as possible, could be used as a resource in estimating breeding values because the markers would pick up echoes of genes affecting quantitative variance for traits, since genes affecting the normal variation are expected to be common and pervasive. So, I chose an extremely polymorphic DNA marker for each chromosome and tested it for 'echoes' to marbling score. The tenth marker, *BM315*, gave an interesting but non-significant result and further testing on that chromosome lead to the first statistically significant result, for the DNA marker *ETH10*. *ETH10* was expected to be close to the gene *IGF1* (insulin-like growth factor 1) due to previous gene mapping experiments and *IGF1* is known to affect lean growth (Ailhaud *et al.*, 1992; Smas and Sul, 1995). I had attended a seminar in 1993 on the inverse gene expression of *IGF1* and *c-MYC* (*v-myc* avian myelocytomatosis viral oncogene homologue) in response to changes in dietary protein (later published as Kanamoto *et al.*, 1994) and since *c-MYC* is located on bovine chromosome 14 (Barendse *et al.*, 1994), we tested *CSSM66* and *RM180*, the former of which also showed statistically significant associations to marbling (marker number 12). All other chromosomes were tested in this way.

We had identified two DNA markers for marbling using a heuristic method testing 44 DNA markers, but then we had to confirm that they were robust and could be found in other breeds. Indeed, we wished to go beyond confirmation to determine whether these DNA markers, or genes close by on the gene map, could act as diagnostic tests for marbling. We chose animals from Angus and Shorthorn families proven to have high marbling scores in some offspring. We compared the marbling scores to DNA marker genotypes and used several markers in each of the two genomic regions to determine whether there were confirmed associations with marbling scores. On one of the chromosomes the search was easy, but on the other it was a case of 'feeling about in the dark'.

As before, we used heuristic methods to determine which gene to test next. For chromosome 14, the testing of TG (thyroglobulin) was partly a case of serendipity; I knew that *MYC* was difficult to work with and the only other gene known at the time to be near it and in the target region on the bovine map was TG. I expected it also to be difficult to work with, due to a known lack of polymorphism dating back to when Georges had tested it as a candidate for double muscling (Baas *et al.*, 1984; Georges *et al.*, 1987), but it had bovine sequence so I felt that if it showed an association then we might want to pursue *MYC* further. After several attempts I was able to get a good polymorphism for TG (*TG5*) and that did show an association, but by then TG had turned into a candidate gene in its own right, after I appreciated the relationship between the action of the thyroid hormones and fat cell growth and differentiation. To this day, none of the other DNA markers tested on chromosome 14 has so far shown an association to marbling when tested as a population association. For chromosome 5 the search for the causative genes continues, even though we have diagnostic DNA markers. In the confirmation phase *ETH10* did not stand up to scrutiny. Neither did *IGF1* (incidentally, this removed the importance of *MYC* on chromosome 14) and more markers were tested, expanding out in both directions. Two markers some 20 cM proximal to *ETH10* did show population associations. These are *LALBA* and *CSSM34*, with the *CSSM34* association being by far the stronger. This implied either a different genetic basis in the Angus and Shorthorn cattle or some other more complex model. We then took this information and tested the region from *CSSM34* to *ETH10* in a group of randomly chosen cattle, primarily of the Angus and Shorthorn breeds. We were surprised to find two associations, one to *CSSM34* and one to markers close to

ETH10. DNA markers in the middle do not show associations to marbling, providing evidence that there are two genes for marbling on this chromosome. This promises to be interesting but identifying the genes is likely to be difficult.

The summary of these experiments is that four DNA markers were found that were consistently associated with differences in marbling score. Three of these DNA markers were anonymous DNA microsatellites (STRP, simple tandem repeat polymorphisms) and one was a SNP (single/simple nucleotide polymorphism in TG). Two of the STRPs used in these experiments were located to YAC clones that contained the genes *RARG* (*CSSM34*) and *RDH5* (*ETH10*) respectively. *RARG* is the retinoic acid receptor gamma, retinoic acid being a physiologically active derivative of vitamin A (retinol), known to have profound effects during the development of animals. Indeed, levels of vitamin A are known to be associated with differences in marbling score in Japanese cattle. *RDH5* is retinol dehydrogenase 5 which is expressed in the eye; *RDH5* is thus unlikely to be a marbling gene but is more likely a good indicator of where the true marbling gene is likely to be in the genome. This is consistent with the variable population association results we found for *ETH10*. The candidate genes that are currently being used or investigated are thus in biochemical pathways known to affect marbling score, fat deposition and the growth and differentiation of adipocytes; these include pathways such as the thyroid/steroid hormone axis and the retinal/retinoid conversion pathways and receptors (Salter, 1950; Burroughs *et al.*, 1958; Ailhaud *et al.*, 1992; Darimont *et al.*, 1993; Smas and Sul, 1995; Torii *et al.*, 1996; Mears *et al.*, 2001).

Reasons for Patenting the Invention

DNA markers can be used to diagnose particular states of the animal and so they may be offered in commercial testing. While the diagnosis could in theory be done by anyone who was appropriately trained, the infrastructure to provide and validate many tests is expensive. The tests provide revenue to companies but those companies need to make a profit or be subsidized in some way. During the last two decades, much DNA technology has been protected and supports the biotechnology industry. So while subsidies might work in certain circumstances, the power of market forces is well known, and allowing someone to make a profit seemed a surer way of ensuring that the tests were available to the public.

When the tests were first described, the immediate response was that they should be protected by patenting since this would allow for exclusivity of commercialization. The alternative, a trade secret, was more dangerous since the likelihood of a trade secret of this nature surviving over many years is low, i.e. someone else might file a patent first or the secret might be revealed. Protection as a trade secret is also less rewarding for the scientists, since publication is obviously not an option in this situation.

The decision to patent occurred as soon as the *TG5* polymorphism was shown to have a population association to marbling scores, since it was then already appreciated that the thyroid hormones affect adipocyte growth and differentiation. As further information on the other markers became available, these were also added to the specification. At first there was some concern when *ETH10* failed to be replicated, but once *CSSM34* was confirmed in two independent experiments, which meant one could be sure of a chromosome 5 gene for marbling, it was added.

Patenting the use of tests for genes that are part of the normal variation is difficult and much thought was given to understand what this process could actually achieve. Firstly, there is the technical issue of whether the tests could be accurately specified, since there is always a certain vagueness to be expected if other genes and the environment also affect the trait. Following on from this, if the test cannot be accurately specified, and by this I mean that the causal DNA change is not specified, even though the test works efficiently, then it is possible that others may stumble upon the true DNA change and patent it (leaving aside for the moment how they would know it was the true DNA change), or worse, use other DNA markers in linkage disequilibrium to the patented DNA marker to offer tests which do not infringe the patent. This pair of problems is rare when the causal mutations for congenital diseases or deleterious mutations are discussed, because in general the mutation makes sense: there is a stop codon, an exon is removed, an intron is left in the mRNA, there is a massive deletion of the coding sequence, there is an amino acid change which can be linked clearly to the functional change, some such clear difference in the DNA. If there is genetic heterogeneity each of the additional mutations also makes sense. If at first one cannot find this sort of genetic change one searches until it is found. However, it is thought that much of the normal variation will not be due to such differences in DNA, but may be due to gene regulation or amino acid changes which give rise to subtle differences in the protein's activity, in which case many of the genetic differences will be in non-coding regions of the DNA; this would make it extremely difficult to determine whether the differences are likely to be causal or merely in linkage disequilibrium. The main point in all of this, in patenting and commercialization, is that several polymorphisms will be expected to be of similar utility to the causal mutation in predicting marbling score and, unless one found all of them, one could question the worth of patenting the test. We came to the conclusion that it was worthwhile, since those who would circumvent the test by using other polymorphisms in the same gene would need to carry out costly research to show which of their alleles was favourable for marbling and still face the risk of legal action alleging that their test infringes our patent.

During the patenting process, several interesting or important issues arose; of course, our due diligence had given us confidence that no one else had filed data for a similar patent, so we could attempt to patent the information that we had. During the preliminary international examination, several of the claims were considered not to be inventive. These were claims that might be interpreted to mean that the methods of detecting the polymorphism had been invented and that specific methods used to visualize the TG polymorphisms had been invented. In addition, due to prior sequencing in humans and mice, the bovine retinoic acid receptor sequence was not considered to be novel or inventive. Nevertheless, this left all claims of utility intact. During the patenting process, US Patent 5,612,179 (Simons, 1997) came to light; this patent has a claim to the use of any DNA test based on an intronic sequence, and here intronic is meant in the broadest possible way to include inter-genic and so-called junk DNA sequences, to predict the genotype at a 'non-intronic' sequence. This surprising patent uses the historically well-known principle of allelic association (Robbins, 1918; Crow and Kimura, 1970; Edwards, 1980) that occurs between loci that are close together on the same chromosome. This patent can be applied in a limited way also to parentage assignment where 'intronic' DNA sequences associated with the human leukocyte antigen (HLA) complex are used as surrogates for HLA tests. This did not invalidate the claims of the marbling patent application but it might have meant that a licence would have been needed if the *TG5* test was being used to predict genotypes at coding sequence polymorphisms of the TG gene. However, *TG5* is used to predict aspects of the physical characteristics of the animal. Finally, the US Patent Office took a more stringent attitude than other patent offices, not only to specific claims but also to what can be incorporated into a single patent: the claims were made to be as specific as possible and were restricted purely to marbling in cattle and each genetic region had to be considered independently since there was no necessary link demonstrated between the genes. Further, alleles that are in linkage disequilibrium that might predict marbling, could not be covered by blanket statements of principle if data for linkage disequilibrium had not been collected for those polymorphisms. Nevertheless, a US patent has been issued for the *TG5* polymorphism.

Research Implications

Not all of the DNA tests are for polymorphisms in genes; three of the tests are anonymous DNA markers, one of which has been associated with a reasonable candidate, *RARG* (see above). These anonymous DNA markers have many genotypes; typically 30 or 40 are possible, most of which may not be seen in a small sample, not just three as is found with

SNPs. This presents problems for prediction of genetic merit, since few animals could be found for some of the genotypes even in large studies of a thousand or more samples. This would make the estimation of merit imprecise. Either we would need to collect even larger samples, or generate and evaluate SNP markers for candidate genes near the anonymous DNA markers. We are doing the latter as this offers the chance that we will obtain a better understanding of the growth and development of marbling by identifying genes affecting marbling.

Identifying the genes that cause the trait is only one step towards our understanding of the genetic basis of normal variation. Identifying the causal mutations for QTL on a broad scale for many traits is a more daunting task. One should expect that it would be possible with sufficient study to identify the gene in which the causal mutation resides, not merely by continued and intensive genetic analysis of genes in a very narrow region of the genome, but also by studying the function, coordination and expression of genes. However, it is likely that no amino acid differences will be found. For example, the differences found between teosinte and maize at the phenotypic level are immense but they have a great deal of similarity at the amino acid level. It is when the untranscribed regulatory regions are studied that major differences can be uncovered at genes important in explaining the differences between these plants (Wang *et al.*, 1999). Novel ideas will be needed to explain these sorts of DNA differences; since they are not transcribed, these differences cannot, in general, be addressed using standard tools. For example, while the thyroid hormones have been linked to marbling and adipocyte growth and development, they are merely contained in the thyroglobulin molecule and produced during the breakdown of TG under a strong negative feedback control. This negative feedback, and the large store of thyroid hormones in the thyroid gland (McNabb, 1992; Williams, 1994), would negate any minor coding sequence change that merely produced fewer thyroid hormone molecules for each TG molecule, while major disruptions of the coding sequence leads to goitre or other deleterious disruptions of thyroid function; TG appears to have low variability in the DNA sequence both in humans and cattle consistent with a DNA sequence under strong selection. The *TG5* polymorphism is in the 5′ leader sequence and the SNP is in a consensus binding sequence for RNA polymerase III (Galli *et al.*, 1981; Ciliberto *et al.*, 1982; Watanabe *et al.*, 1982; Skowronski *et al.*, 1984). We do not know whether this is important and it is not yet clear how one would go about determining whether this difference was causal.

These DNA markers have been evaluated in several environments in Australia but to have general utility they need to be evaluated in as many different countries and feeding systems as possible. The contribution of particular genes to phenotypic variation has been shown to depend on the environment and circumstances in which the organism grows (Paterson *et al.*, 1991) and it is presumed that the same applies to marbling. For example, in one set of experiments for the TG5 marker we found the heterozygous animals to be intermediate in genetic potential to the two homozygotes, but in a different set of experiments the heterozygote appeared indistinguishable from one of the homozygotes. Such heterogeneity of effect should be expected in other systems as well, so a general rule would be that once these markers have been found they need to be evaluated in as many systems as possible.

While the markers may be used in isolation they will be more beneficial when used in conjunction with other DNA markers and with evaluation of breeding values for marbling scores. One of the major misunderstandings of this kind of technology is that variation at a single gene is responsible for essentially all of the genetic variation that is observed; gene influence is erroneously referred to as gene control. While the effects of the genes are moderate, they are clearly influenced by the effects of other genes and the environment, otherwise there would be clear bimodality in the distributions of marbling score. So if other genes are important, the converse is that improved prediction will result as other genes are included in the test. We could test this explicitly with new samples, re-analysing the existing data has the flaw that such analyses are *a posteriori* so we expect to see enhancements (and we do). However, we can go beyond adding other DNA markers to our predictions; we can use phenotypic measurements of marbling performance. Unfortunately, this is much easier said than done, since there appears to be no strong consensus on the best way in which this should be done, nor a good explanation for the unpredictable way in which sires are re-ranked as one adds molecular information when estimating breeding values.

Responses from the Public and Researchers and a Cautionary Tale

Once patent applications were filed for the DNA markers, expressions of interest were invited for their commercialization, following which they were licensed to Genetic Solutions Pty Limited. The existence of DNA markers for marbling was kept under wraps for a long time. While the allelic associations had been detected in 1997 and marbling QTL were first described in the lab in 1995, the path to market was not laid out nor had a commercializer been chosen, so broadcasting the existence of such a test would merely have raised public pressure for a test with consequent disappointment when the test was not offered in the following months. The search for a licensee began in 1998 with a small number of offers for the technology, partly because the technology was new and few companies knew how to implement it properly. These

DNA markers were offered as part of a package of technologies and would be commercial partners were invited to submit proposals not only for what they might do with these diagnostic DNA markers but also what they would do with DNA markers that were linked to other traits. Ultimately Genetic Solutions was chosen as the commercial partner because this company had the most integrated plan on what to do with all the markers, part of which involved plans for future research. The flagship product of that commercialization is the TG5 test marketed as GeneSTAR Marbling. Now that the markers have been commercialized and other DNA markers are appearing for similar traits, we receive many more offers to commercialize new technology.

Thousands of tests have now been sold in the 18 months that they have been available. Publicity for the tests was low initially as the commercializer geared up for testing; a media release was organized by CSIRO for just after New Year 2001, to link the usual over-consumption of the Christmas festivities and concerns about diet and obesity in humans with a story about genes affecting fatness in cattle having possible links in humans. This story attracted international attention from news services as well as domestic print, radio and television. Such attention might publicize the test, and care was taken to ensure that the real story did not get lost in speculation about human fatness, but the test itself and its utility must of necessity be marketed at the ground level to cattle producers, an activity pursued by the commercializers. The thousands of tests done are an indication of the importance of marbling to producers and the continuing emphasis on genetic tools for animal production.

Nevertheless, there is still some confusion about what the test does and how it will be used over the long term. Part of the confusion results from the usual expectation that a single genetic test for marbling will act like a silver bullet, useful to breed an animal of high marbling performance from ancestors of low marbling performance in a single generation. This cannot occur due to the complex inheritance of marbling potential and breeding schemes need to take into consideration not just the *TG5* results but also results from other DNA markers and the breeding values for marbling score, where these are available. Providing access to such other data will occur over time; other markers will be offered as part of the DNA test while rough and ready breeding values for intramuscular fat, based primarily on ultrasound scanning, are now also becoming available (e.g. Sundstrom *et al.*, 1999). Attempts are being made to integrate the *TG5* data into estimated breeding values for marbling so that both sorts of information can be provided to producers as a consistent tool, to reduce confusion and provide greater predictability.

The integration of all data on marbling is an important goal, and the lack of such integration may have slowed the implementation of previous attempts by other groups to use predictive molecular markers for a production trait. Several decades ago the kappa casein protein polymorphism was found to be associated with milk protein yield (e.g. McLean *et al.*, 1984; Gibson, 1990). This protein polymorphism was identified before patenting of such material was a routine operation, so anyone was free to do the test. After an initial burst of enthusiasm it was found that the best bulls for milk protein yield need not necessarily have the favourable genotype, and there was no expectation that all the best bulls would have the favourable genotype. The casein marker then fell into disuse for many years, partly perhaps because its other advantage for predicting cheese quality was not seen as important at the time and partly because milk protein yield is measured extremely accurately. Perhaps if the casein marker had been integrated into breeding value estimation it might not have fallen out of favour; but it is also likely that the expectations of producers were that protein yield could be predictable with a single test. Nevertheless, there has been a resurgence in the use of this casein marker, now formatted as a DNA test, and breeds advertise high frequencies of the favourable form on the World Wide Web. In contrast, the *TG5* marker is patented and licensed exclusively to one company which is intimately involved in attempts to integrate the molecular information into breeding value estimates. In this way we hope to avoid the fate of the casein marker – spending years in the wilderness.

More of the DNA markers are being produced by ourselves and by other research groups offering promise of new techniques but new responsibilities. The DNA markers will be for traits like meat tenderness, feed conversion efficiency and parasite resistance, which are difficult to improve via conventional methods, either because the trait is difficult or expensive to measure accurately or because of low heritability (e.g. fertility), so conventional breeding will show little progress over many generations. Such research is also at the forefront of the study of pharmacogenomics and of the risk factors for disease in humans, in efforts to develop tests that predict genetic tendencies in response to the environment, medicine and treatment. In this kind of work, the tests will be useful but they must be integrated into complete strategies. Failure to do so will undoubtedly lead to a proportion of incorrect decisions being made. While this might be of minor inconvenience in livestock, it may have far reaching ramifications in human health, if diagnoses are made purely upon the basis of DNA tests for QTL for tendencies to a particular response.

Conclusions

The understanding of the genetics of animals and plants has reached the point where we are beginning to identify genes that contribute to the normal genetic variation or what is often referred to as the genetic potential for characteristics. We have thus moved beyond genes that have large or congenital effects on characteristics and are now identifying genes affecting characteristics as diverse as hypertension or cancer in humans and meat quality and composition in livestock. These newer genetic tools can be patented and commercialized in a similar way to DNA tests for large mutations; however, they are more difficult to protect and to commercialize since their effects are mitigated by the influence of other genes and the environment. This paper has described how markers for marbling in beef were identified, why they were patented, how they are currently being offered to the public and what the tests may mean for the beef industry.

Acknowledgements

Greg Davies, Peter Willadsen and Shaun Coffey made comments which improved the manuscript. The research on which this account is based was funded in part by MLA (Meat and Livestock Australia).

References

Ailhaud, G., Grimaldi, P. and Negrel, R. (1992) Cellular and molecular aspects of adipose tissue development. *Annual Reviews of Nutrition* 12, 207–233.

Antonarakis, S.E. (1994) Genome linkage scanning – systematic or intelligent. *Nature Genetics* 8, 211–212.

AUS-MEAT (1998) *Chiller Assessment Progam*. Standard Manual, 3rd edition.

Baas, F., Bikker, H., van Ommen, G.J. and de Vijlder, J.J. (1984) Unusual scarcity of restriction site polymorphism in the human thyroglobulin gene. A linkage study suggesting autosomal dominance of a defective thyroglobulin allele. *Human Genetics* 67, 301–305.

Barendse, W. (1997) Assessing lipid metabolism. Patent WO9923248 (Patent Application PCT/AU98/00882). <http://ep.espacenet.com/>

Barendse, W., Armitage, S.M., Kossarek, L.M., Shalom, A., Kirkpatrick, B.W., Ryan, A.M., Clayton, D., Li, L., Neibergs, H.L., Zhang, N., Grosse, W.M., Weiss, J., Creighton, P., McCarthy, F., Ron, M., Teale, A.J., Fries, R., McGraw, R.A., Moore, S.S., Georges, M., Soller, M., Womack, J.E. and Hetzel, D.J.S. (1994) A genetic linkage map of the bovine genome. *Nature Genetics* 6, 227–235.

Brethour, J.R. (2000) Using serial ultrasound measures to generate models of marbling and backfat thickness changes in feedlot cattle. *Journal of Animal Science* 78, 2055–2061.

Burroughs, W., Raun, A. and Cheng, E. (1958) Effects of methimazole on thyroid and live weights of cattle. *Science* 128, 147.

Ciliberto, G., Traboni, C. and Cortese, R. (1982) Relationship between the two components of the split promoter of eukaryotic tRNA genes. *Proceedings of the National Academy of Sciences USA* 79, 1921–1925.

Coleman, J.B., Cucca, F., Hearne, C.M., Cornall, R.J., Reed, P.W., Ronningen, K.S., Undlien, D.E., Nistico, L., Buzzetti, R., Tosi, R., Pociot, F., Nerup, J., Cornelis, F., Barnett, A.H., Bain, S.C. and Todd, J.A. (1995) Linkage disequilibrium mapping of a type 1 diabetes susceptibility gene (IDDM7) to chromosome 2q31–q33. *Nature Genetics* 9, 80–85.

Collins, F.S. (1995) Positional cloning moves from perditional to traditional. *Nature Genetics* 9, 347–350.

Crow, J.F. and Kimura, M. (1970) *An Introduction to Population Genetics Theory.* Harper and Row, New York.

Darimont, C., Gaillard, D., Ailhaud, G. and Negrel, R. (1993) Terminal differentiation of mouse preadipocyte cells: adipogenic and antimitogenic role of triiodothyronine. *Molecular and Cellular Endocrinology* 98, 67–73.

Dekkers, J.C.M. and Dentine, M.R. (1991) Quantitative genetic variance associated with chromosomal markers in segregating populations. *Theoretical and Applied Genetics* 81, 212–220.

Edwards, J.H. (1980) Allelic association in man. In: Eriksson, A.W., Forsius, H.R., Nevanlinna, H.R., Workman, P.L. and Norio, R.K. (eds) *Population Structure and Genetic Disorders.* Academic Press, London, pp. 239–255.

Galli, G., Hofstetter, H. and Birnstiel, M.L. (1981) Two conserved sequence blocks within eukaryotic tRNA genes are major promoter elements. *Nature* 294, 626–631.

Georges, M., Lequarre, S., Hanset, R. and Vassart, G. (1987) Genetic variation of the bovine thyroglobulin gene studied at the DNA level. *Animal Genetics* 18, 41–50.

Gibson, J.P. (1990) Is there profit in a protein gene? *Holstein Journal*, December. <http://apsit.aps.uoguelph.ca/pub/articles/holstjg2.html/>

Grobet, L., Martin, L.J.R., Poncelet, D., Pirottin, D., Browers, B., Riquet, J., Schoeberlein, A., Dunner, S., Menissier, F., Massabanda, J., Fries, R., Hanset, R. and Georges, M. (1997) A deletion in the bovine myostatin gene causes the double-muscled phenotype in cattle. *Nature Genetics* 17, 71–74.

Hanset, R., Michaux, C., Dessy-Doize, C. and Burtonboy, G. (1982) Studies on the 7th rib cut in double muscled and conventional cattle. Anatomical, histological and biochemical aspects. In: King, J.W.B. and Menissier, F. (eds) *Muscle Hypertrophy of Genetic Origin and its Use to Improve Beef Production.* Martinus Nijhoff, The Hague.

IHGSC (2001) International Human Genome Sequencing Consortium. Initial sequencing and analysis of the human genome. *Nature* 409, 860–921.

Kanamoto, R., Yokota, T. and Hayashi, S.I. (1994) Expressions of c-MYC and insulin-like growth-factor-1 messenger-RNA in the liver of growing rats vary reciprocally in response to changes in dietary-protein. *Journal of Nutrition* 124, 2329–2334.

Lembertas, A.V., Perusse, L., Chagnon, Y.C., Fisler, J.S., Warden, C.H., Purcell-Huynh, D.A., Dionne, F.T., Gagnon, J., Nadeau, A., Lusis, A.J. and Bouchard, C. (1997) Identification of an obesity quantitative trait locus on mouse chromosome 2 and evidence of linkage to body fat and insulin on the human homologous region 20q. *Journal of Clinical Investigations* 100, 1240–1247.

Lynch, H.T. and Watson, P. (1992) Genetic counselling and hereditary breast/ovarian cancer. (Letter) *Lancet* 339, 1181.

McLean, D.M., Graham, E.R.B., Porzoni, R.W. and McKenzie, H.A. (1984) Effects of milk protein genetic variants on milk yield and composition. *Journal of Dairy Research* 51, 531–546.

McNabb, F.M.A. (1992) *Thyroid Hormones*. Prentice Hall, Englewood Cliffs, New Jersey.

Mears, G.J., Mir, P.S., Bailey, D.R.C. and Jones, S.D.M. (2001) Effect of Wagyu genetics on marbling, backfat and circulating hormones in cattle. *Canadian Journal of Animal Science* 81, 65–73.

MSA (2002) Meat Standards Australia. Accessed via http://www.mla.com.au/

OMIA (2002) Online Mendelian Inheritance in Animals. http://www.angis.org.au/Databases/BIRX/omia/

OMIM (2002) Online Mendelian Inheritance in Man. http://www.ncbi.nlm.nih.gov/

Paterson, A.H., Lander, E.S., Hewitt, J.D., Peterson, S., Lincoln, S.E. and Tanksley, S.D. (1988) Resolution of quantitative traits into Mendelian factors by using a complete RFLP linkage map. *Nature* 335, 721–726.

Paterson, A.H., Damon, S., Hewitt, J.D., Zamir, D., Rabinowitch, H.D., Lincoln, S.E., Lander, E.S. and Tanksley, S.D. (1991) Mendelian factors underlying quantitative traits in tomato – comparison across species, generations, and environments. *Genetics* 127, 181–197.

Robbins, R.B. (1918) Some applications of mathematics to breeding problems. *Genetics* 3, 375–389.

Salter, W.T. (1950) The chemistry and physiology of the thyroid hormone. In: Pincus, G. and Thimann, K.V. (eds) *The Hormones, Physiology, Chemistry and Applications*. Academic Press, New York, pp. 181–299.

Savell, J.W., Cross, H.R. and Smith, G.C. (1986) Percent ether extractable fat and moisture content of beef longissimus muscle as related to USDA marbling score. *Journal of Food Science* 51, 838–840.

Simons, M.J. (1997) Intron sequence analysis method for detection of adjacent and remote locus alleles as haplotypes. US Patent 5,612,179.

Skowronski, J., Plucienniczak, A., Bednarek, A. and Jaworski, J. (1984) Bovine 1.709 satellite recombination hotspots and dispersed repeated sequences. *Journal of Molecular Biology* 177, 399–416.

Smas, C.M. and Sul, H.S. (1995) Control of adipocyte differentiation. *Biochemical Journal* 309, 697–710.

Sokal, R.R. and Rohlf, F.J. (1981) *Biometry*, 2nd edn. W.H. Freeman & Co., San Francisco.

Soller, M., Brody, T. and Genizi, A. (1976) On the power of experimental designs for the detection of linkage between marker loci and quantitative loci in crosses between inbred lines. *Theoretical and Applied Genetics* 47, 35–39.

Sundstrom, B. (1999) The new carcass EBVs, version 4.1. http://breedplan.une.edu.au/bnotes/bnotes.htm

Thoday, J.M. (1961) Location of polygenes. *Nature* 191, 368–370.

Todd, J.A., Bell, J.I. and McDevitt, H.O. (1987) HLA-DQ(beta) gene contributes to susceptibility and resistance to insulin-dependent diabetes mellitus. *Nature* 329, 599–604.

Torii, S., Matsui, T. and Yano, H. (1996) Development of intramuscular fat in Wagyu beef cattle depends on adipogenic substances present in serum. *Animal Science* 63, 73–78.

Wang, R.L., Stec, A., Hey, J., Lukens, L. and Doebley, J. (1999) The limits of selection during maize domestication. *Nature* 398, 236–239.
Watanabe, Y., Tsukada, T., Notake, M., Nakanishi, S. and Numa, S. (1982) Structural analysis of repetitive DNA sequences in the bovine corticotropic-b-lipotropin precursor gene region. *Nucleic Acids Research* 10, 1459–1469.
Williams, G.R. (1994) *Thyroid Hormone Regulation of Gene Expression*. R.G. Landes Co., Molecular Biology Intelligence Unit, Austin, Texas.
Woolf, B. (1955) On estimating the relation between blood group and disease. *Annals of Human Genetics* 19, 251–253.
Worrell, M.A., Clanton, D.C. and Calkins, C.R. (1987) Effect of weight at castration on steer performance in the feedlot. *Journal of Animal Science* 64, 343–347.

13

A Research and Patent Perspective of Nuclear Transfer Cloning: Case Studies

Robin Teskin[1] and José Cibelli[2]

[1]*Pillsbury, Madison & Sutro, LLP, 1100 New York Avenue, N.W. Ninth Floor, East Tower, Washington, DC 20005-3918, USA;* [2]*Advanced Cell Technologies, One Innovation Drive, Biotech Three, Worcester, MA 01605, USA*

Introduction

The line of experimentation that eventually led to the development of the concepts and procedures of nuclear transfer cloning derived from interest in questions about the totipotency of committed (differentiated) somatic cells. The dominant hypothesis at the beginning of the 20th century was that following fertilization the nucleus became more and more restricted in its developmental potential (Spemann, 1914). Successive experiments gradually weakened this view until Wilmut *et al.* (1997) showed that a lamb could be cloned from a fully differentiated adult cell nucleus of a cultured mammary epithelia from a 6-year-old ewe. Early on, Spemann (1914) showed that a piece of the fertilized newt egg without a nucleus could be induced to develop by the introduction of a descendant of the fertilized egg nucleus (Spemann, 1914, 1938). Using a hair loop, he ligated the egg so that only one portion contained the nucleus. Cell cleavage (8–16 cells) was restricted to that portion. After a few rounds of cell division, the ligature was released and one cell was permitted to migrate to the other side. Finally the embryo was completely separated into two entities. In some cases each half developed into a larvae, indicating that nuclei from an embryo already committed in development (8–16 cell stage) could still generate a new embryo. The microsurgical procedures for oocyte enucleation, and nuclear injection that became the prototype for nuclear transfer were developed by Briggs and King (1952). They injected an early *Xenopus* gastrula nucleus into the cytosol of an enucleated non-fertilized egg and observed development of fully grown animals.

 Intellectual Property Rights in Animal Breeding and Genetics (eds M. Rothschild and S. Newman)

Thus, nuclei from determined and differentiated cells retained the totipotency of the zygote. Gurdon *et al.* (1975) further extended this experiment by injecting fully differentiated *Xenopus* skin cells into enucleated unfertilized oocytes. Although only tadpoles were generated, it was evident that cells already fully committed to make skin retained the capacity to generate an individual with all of the different tissues.

Efforts to repeat these results in mammals were soon attempted in the mouse with controversial results. While Hoppe and Ilmensee (1982) claimed the successful generation of mice from an inner cell mass cell transplanted into an enucleated zygote, these experiments were soon challenged by McGrath and Solter (1983) and the capacity to clone from a mammalian cell remained an uncertainty.

The first accepted successful cloning event (of a sheep) was reported by Willadsen (1986). Using a single blastomere taken from an eight-cell embryo and placed into a non-fertilized recipient egg, Willadsen reported the production of a healthy lamb. Similar experiments were soon repeated in bovine, rabbit and pig. Notwithstanding these results, the experiments with frog somatic cells performed by Gurdon were not attempted in mammals until 1997 when a lamb was cloned from a cell from the mammary gland of an adult sheep (Wilmut *et al.*, 1997). Successful reports in mice, bovine, goat and pig followed (Cibelli *et al.*, 1998; Wakayama *et al.*, 1998; Baguisi *et al.*, 1999; Onishi *et al.*, 2000).

Biological Background

During normal fertilization, the egg and the sperm will contribute the DNA necessary to create the offspring. During cloning however, all maternal DNA from the egg must be removed and replaced by the genomic DNA from the donor animal. This process requires precise microscopic surgery to remove the egg's own chromosomes by a process called enucleation, followed by the introduction of a somatic cell into the periviteline space (between the zona pellucida and the egg's cell membrane) and fusion with the egg with electricity. Alternatively, the nucleus of the donor cell can be directly injected into the cytosol of the egg using a piezo-electric device avoiding the fusion step altogether. After reconstruction, a series of nuclear changes will occur in the egg, the DNA of the new 'zygote' will condense and will remain as such until the onset of embryonic development is artificially triggered. The one cell embryo (zygote) will divide and upon transfer into the uterus of a recipient female, offspring can be generated.

Donor cells: somatic vs. blastomeres (early embryonic cells)

Somatic cells are derived from the body of an animal (Fig. 13.1a). They differ from germ cells or gametes in that they do not have the capacity to generate a new individual via fertilization.

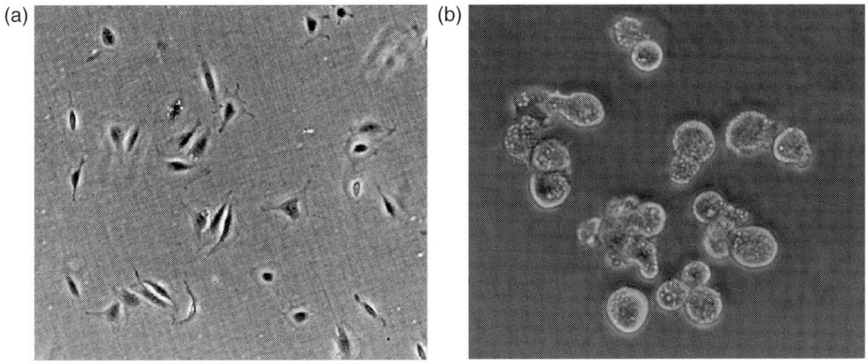

Fig. 13.1. Early embryonic cells. (a) Fibroblasts (body cells); (b) blastomeres (embryonic cells).

Blastomeres, or early embryonic cells, are those isolated from an embryo that has undergone the first rounds of division after fertilization, somewhere from two to 10 divisions depending on the species. As such, blastomeres have the potential to become any cell in the body including the placenta in mammals (Fig. 13.1b). Prior to 1996, the dogma taught that the only cell type capable to give rise to a new individual via cloning was a blastomere. Although this is still remarkable, their potential use was limited. Blastomeres cannot be cultured *in vitro* for a prolonged period of time restricting their uses to exclusive basic-research purposes. From an industrial standpoint, a cell that can be easily cultured *in vitro* can first and foremost become a permanent reservoir of a desired genotype. Further, it has the potential to be genetically modified, opening the possibility to generate animals for biomedical and agricultural applications.

In search for the 'bottle neck' in mammalian cloning

There are a significant number of steps in cloning technology that have been developed over the past 40 years, some of which have been previously discussed. A significant development is the discovery that somatic cells of different cell cycles can be used to clone mammals. This is significant from both a research and economic perspective. It is important from a research perspective, as it is contrary to previous dogma relating to cell totipotency and the supposed inability of the nucleus of a cell committed to a particular lineage to differentiate into all cell lineages. Based thereon, it was widely accepted that the nucleus of a somatic cell would not be capable of giving rise to a cloned mammalian animal. Moreover, it is significant from an economic perspective as somatic cell cloning provides the opportunity to generate limitless numbers of an individual, e.g. an agricultural animal, with a desired genotype. The economic importance of this discovery is reflected by the numerous patent applications that have been submitted relating to nuclear transfer as a direct result of this discovery. By contrast, prior thereto, only a relatively small number of applicants had submitted patents relating to nuclear transfer (this is summarized in more detail in another section of this chapter). Two biological and scientific aspects broke the dogma: (i) the use of a somatic (body) cell as a nucleus donor and (ii) the cell cycle.

In 1997, with the public announcement of the birth of Dolly, somatic (body) cells were introduced as a viable alternative to blastomeres for nuclear transfer experiments. Months later, advanced cell technology (ACT) showed similar results in the cow.

The Roslin group (UK, Dolly creators) had used a technique which at that time was considered to be the 'key' of somatic cell cloning. Fortunately, our group had been working on cloning in parallel at the University of Massachusetts (USA) and such 'key' was never considered nor used in our experiments; this crucial aspect apparently resided on the 'cell cycle' (see Fig. 13.2).

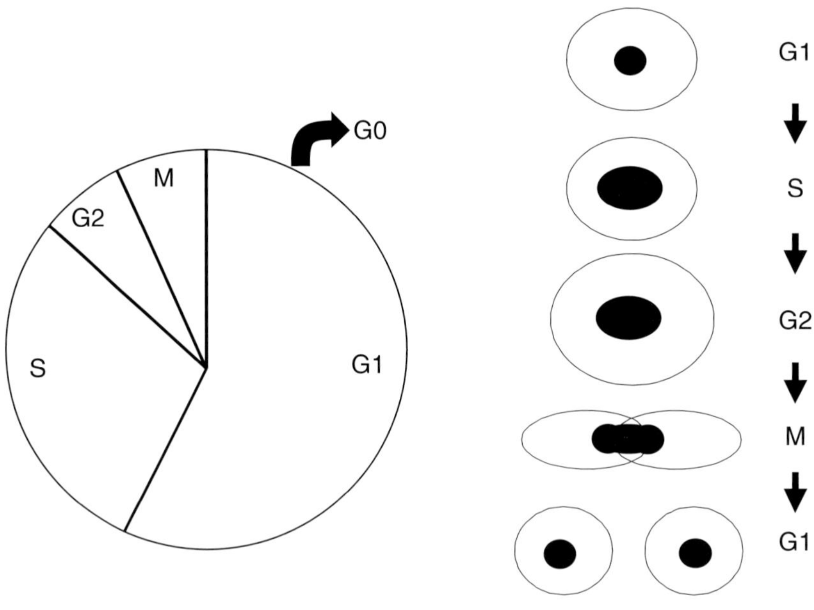

Fig. 13.2. The cell cycle is the process by which a cell grows and divides. Cell division itself, mitosis (M phase) and DNA replication or synthesis (S phase) are the two major events that split the cycle into two gaps (G1 and G2). Cells can be forced out of the cell cycle by a variety of natural and artificial means, among them, removal of their nutrients, DNA damage or terminal differentiation.

The reconstruction of an embryo via nuclear transfer is indeed the merger of a donor cell with a recipient egg. In order to obtain proper embryonic development, a certain degree of synchrony must be in place between both entities otherwise the embryo is halted in its growth. This had been known for more than 10 years (Campbell *et al.*, 1993; Liu *et al.*, 1997). However, Campbell and colleagues were the first to show that cells which have exited the cell cycle (quiescent state or G0) were capable of giving rise to a healthy individual via cloning (Dolly – Wilmut *et al.*, 1997). We at ACT were working on somatic cell nuclear transfer in parallel and although we lost the race to be the first to report the successful cloning of a mammal from a somatic cell, our group was the first to show the cloning of a mammal (bovine) using an *actively dividing* somatic cell. Further studies conducted by independent researchers all over the world have confirmed Roslin and ACT's findings and it is now widely accepted that both actively dividing cells and quiescent cells can be used for somatic cell cloning (Table 13.1). As discussed in the next section this has precipitated the filing of a number of patent applications by Roslin, ACT and many others.

Table 13.1. A review of previous somatic cell cloning research.

Species	Cell cycle	Number of studies	Reference
Mouse	Cycling cells	3	Wakayama and Yanagimachi (1999, 2001); Ono *et al.* (2001)
	Quiescent	3	Wakayama and Yanagimachi (1999, 2001); Yamazaki *et al.* (2001)
	Non-determined	5	Wakayama *et al.* (1998); Zhou *et al.* (2000); Ogura *et al.* (2000a, b); Tamashiro *et al.* (2000)
Sheep	Cycling cells	0	
	Quiescent	3	Denning *et al.* (2001); McCreath *et al.* (2000); Schnieke *et al.* (1997)
	Non-determined	1	Loi *et al.* (2001)
Pig	Cycling cells	2	Bondioli *et al.* (2001); Polejaeva *et al.* (2000)
	Quiescent	1	Onishi *et al.* (2000)
	Non-determined	1	Betthauser *et al.* (2000)
Goat	Cycling cells	1	Baguisi *et al.* (1999)
	Quiescent	3	Reggio *et al.* (2001); Keefer *et al.* (2001); Zou *et al.* (2002)
	Non-determined	0	
Cow	Cycling cells	5	Cibelli *et al.* (1998); Kasinathan *et al.* (2001); Renard *et al.* (1999); Shiga *et al.* (1999); Lanza *et al.* (2001); Zakhartchenko *et al.* (1999, 2001)
	Quiescent	2	Zakhartchenko *et al.* (1999, 2001)
	Non-determined	2	Galli *et al.* (1999); Hill *et al.* (2000)

A Summary Overview of Published Patents and Patent Applications Relating to Nuclear Transfer

Recently there have been tremendous advances in nuclear transfer research. Undeniably the most significant recent discovery is that differentiated cells of different types and different cell cycles, give rise to viable embryos and offspring when used as donor cells for nuclear transfer cloning. This discovery was first reported by Ian Wilmut and Keith Campbell of the Roslin Institute, who disclosed that they had cloned a sheep they named Dolly by nuclear transfer using a quiescent somatic donor cell (Wilmut *et al.*, 1997). The news of Dolly was followed soon after by the research conduct by ACT in collaboration with the University of Massachusetts (ACT group), when we reported that we had cloned three healthy calves by nuclear transfer using somatic donor cells (fetal fibroblasts; Cibelli *et al.*, 1998). In contrast to the Roslin group, the cloning conducted by ACT used proliferating (non-quiescent) donor cells.

The discoveries reported by the Roslin Institute and ACT in collaboration with University of Massachusetts provide an efficient means for producing non-human animals having a desired genotype. This is especially promising in the context of agriculture, as nuclear transfer cloning enables the production of cloned animals having desired characteristics. Also, these discoveries have significant medical applications, e.g. cloned animals can be produced that express a desired protein (e.g. a therapeutic polypeptide), or can be used as a source of cells or tissues having a particular genotype for transplantation therapies. Still further nuclear transfer can be used to produce pluripotent cells and cell lines.

Given the significant agricultural and commercial applications of nuclear transfer, it is perhaps unsurprising that the advances in nuclear transfer research have been accompanied by a significant surge in the number of patent applications filed in the USA and abroad that relate to nuclear transfer. A large number of different universities have filed patent applications relating to nuclear transfer including the Roslin Institute, University of Massachusetts, University of Georgia, University of Montreal, University of Connecticut, University of Monash, University of Singapore, University of Edinburgh and the University of Wisconsin. Companies that have filed patent applications that relate to nuclear transfer include PPL Therapeutics, ACT, Infigen Inc., Geron Corp., Biotransplant Inc., Pangene Corp. and Stem Cell Science Party Limited, among others.

In fact, a computerized search of the published patent literature using as the keywords 'nuclear transfer' revealed that as of January 2002, approximately 150 patent applications have been filed which relate, at least in part, to nuclear transfer. This number does not even take into account the numerous counterpart patent applications that have been filed in other jurisdictions which relate to these applications. It is evident, based on the number of patent applications that have been filed to date, that many companies and universities are seeking to patent their discoveries in the area of nuclear transfer.

As the number of patent applications which have been filed in this area is too voluminous to treat in a single book chapter, it was decided to instead provide an overview of the various types of claims relating to nuclear transfer that are contained in published patent applications or issued patents to date relating to nuclear transfer. For the convenience of the reader, whenever possible, the subject matter of the discussed patent applications or patents is grouped based on related subject matter. Also, the earliest filed patent applications involving nuclear transfer are discussed first, as these applications put into context the state of the art of nuclear transfer prior to the discoveries made by the Roslin Institute and the University of Massachusetts.

Prior to summarizing the numerous published patent applications that relate to nuclear transfer, it should be noted at the outset that the USA, as well as many other countries which grant patents, permits the patenting of methods that provide non-human organisms. In fact, a large number of patents have been granted that are directed to the production of transgenic non-human animals. Also, the USA, as well as some other countries, have patent statutes that permit the patenting of non-human animals produced by recombinant methods provided that the animal is novel and non-obvious to a skilled person as of the date of invention. Similarly, many patents have been granted containing claims that cover transgenic animals. Accordingly, nuclear transfer cloning and the products obtained thereby constitute subject matter, which is potentially eligible for patent protection.

Summary of published patents and patent applications relating to nuclear transfer prior to Dolly

During the early to mid-1990s a number of patent applications were filed, some of which have matured into patents, that claim alleged innovations and improvements in nuclear transfer procedures and materials for use in such procedures. In general, these early patent applications all relate to nuclear transfer/cloning procedures that utilize as the donor cell an undifferentiated embryonic cell or nucleus derived therefrom.

For example, US Patent 5,453,366, based on an application filed on 31 January, 1990 by Rosenkrans and Sims, now assigned to Infigen Inc., is exemplary of such applications. This patent relates to a method of cloning bovine embryos using as the donor cell an embryonic cell (blastomere) derived from a 2–64 cell embryo. Exemplary claim 1 of the patent is reproduced below:

1. A method for producing cloned bovine embryos comprising the following steps:
   a. Removing the chromosomal material from a metaphase II stage bovine oocyte to form an enucleated recipient oocyte;
   b. Maintaining the enucleated recipient oocyte in CR-1+1 mM L-glutamine maintenance medium wherein the maintenance period is approximately 6 to 18 hours;
   c. Placing a membrane bound nucleus from a donor 2–64 cell bovine embryo adjacent to the plasma membrane of the enucleated recipient oocyte and;
   d. Inducing cell fusion between the membranes of the donor membrane-bound nucleus and the enucleated recipient oocyte to form an embryonic single cell with a nucleus from the donor, wherein the membranes are either electrically fused or fused with polyethylene glycol.

Additionally, various patent applications also submitted around the same time describe purported innovations in the procedure for activating an embryo or oocyte so that it is capable of giving rise to a viable offspring. For example, US Patents 5,496,720, 5,843,754, 6,194,202 and 6,077,710, which name Susko-Parrish *et al.* as the inventors and assigned to Infigen Inc., are all based on a US patent application submitted in early 1993 and relate to a parthenogenic oocyte activation procedure, whereby intracellular levels of divalent cations in the oocyte are increased and the phosphorylation of cellular proteins in the oocyte are decreased. The granted claims of the 5,496,720, 5,843,754 and 6,077,710 patents are limited to activation of bovine oocytes. By contrast, the most recently issued patent, US Patent 6,077,710, claims a parthenogenic activation of 'mammalian oocyte', as shown by claim 1 of the '710 patent reproduced below:

1. A process for the *in vitro* parthenogenic activation of a mammalian oocyte comprising the following steps:
   a. Increasing intracellular levels of divalent cations in the oocyte; and
   b. Reducing phosphorylation of cellular proteins in the oocyte.

Another nuclear transfer patent application filed around this time, now US Patent 6,107,543, is based on a patent application by Neal First and Michelle Sims and is directed to a method for cloning bovine animals by nuclear transfer, again using embryonic donor cells. As reflected by claim 1 of the '543 patent reproduced below, the donor cell utilized in the procedure patented by First and Sims is a cultured inner cell mass cell that 'has been cultured in a culture medium for a length of time sufficient for maintenance of said inner cell mass cell in an undifferentiated state'.

1. A method for producing a bovine animal, the method comprising:
   a. Providing a cultured bovine embryonic inner cell mass cell, wherein said inner cell mass cell has been cultured in a culture medium for a length of time sufficient for maintenance of said inner cell mass cell in an undifferentiated state;
   b. Establishing an embryo from a nuclear transfer process, wherein said nuclear transfer process comprises the step of inserting said cultured bovine embryonic inner cell mass cell into an enucleated bovine oocyte, and the step of activating said embryo;
   c. Implanting said embryo into a recipient bovine host; and
   d. Allowing said embryo to develop into the bovine animal.

The patent further describes that the inner cell mass cells are cultured in loose suspensions so as to provide for the retrieval of large numbers of stem cells.

Another patent application relating to the use of embryonic cells as donor cells for nuclear transfer is GB9417831, filed 5 September 1994 by the Roslin Institute. This application, by Keith Campbell *et al.* is directed to nuclear transfer cloning methods that use a donor cell derived from the embryonic disc of an ungulate blastodermic vesicle. This application has not been granted in any jurisdiction.

Also submitted around the same time were patent applications directed to culture media and stabilization procedures for maintaining nuclear transfer bovine embryos. For example, US Patent 5,160,312 based on a patent application filed by Steven Voelkel in 1990 patents a method for cryopreservation of bovine embryos using polyethylene glycol. Also, US Patent 5,096,822 based on a patent application by Carol L. Keefer and others, patents a bovine embryo culture medium containing hemicalcium lactate.

Published patent applications directed to the use of non-embryonic cells as nuclear transfer donor cells

Beginning in 1995, a number of applicants starting with the cloners of Dolly submitted patent applications relating to nuclear transfer procedures that utilize non-embryonic donor cells (differentiated donor cells). Applicants who have filed patent applications relating to the use of non-embryonic donor cells include, e.g. the Roslin Institute, the University of Massachusetts (which are exclusively licensed to ACT), Infigen Inc. and PPL Therapeutics.

For the reasons set forth previously, the methods which are the subject of these patent applications constitute a significant advance over earlier nuclear transfer procedures which utilized embryonic donor cells. For example, non-embryonic cells (somatic cells) can be more easily genetically modified in culture than embryonic cells, thereby facilitating the production of genetically modified clones. Some of those applications are summarized below:

Roslin Institute patent applications

On 31 August 1995, the Roslin Institute filed two patent applications in the UK describing nuclear transfer procedures that encompass the use of non-embryonic donor cells.

The first application, GB9517780, by Ian Wilmut and Keith Campbell, relates to the nuclear transfer procedure that was used to clone Dolly. This application describes a nuclear transfer procedure that includes the transferral of a *quiescent* diploid donor cell into a suitable recipient cell. The patent application discloses that the recipient cell is typically an oocyte and that the donor diploid quiescent cell includes differentiated cells, adult somatic cells, fetal cells and embryo donor cells. The patent application further states that the donor cell can be genetically modified.

The European and US counterpart of this UK application have been granted as EP 849 990 and US Patent 6,147,270 respectively. Both of these patents were granted with similar method claims directed to producing reconstituted non-human embryos. Whereas the claims of the European patent provide for the reconstituted embryo to be 'a non-human animal embryo', the US patent is limited to 'non-human mammalian embryo'. Claim 1 of both patents is set forth below for comparison:

Claim 1 EP 849 990

1. A method of reconstituting a non-human animal embryo, the method comprising transferring the nucleus of a quiescent diploid donor cell into a suitable recipient cell.

Claim 1 US Patent 6,142,276

1. A method of reconstituting a non-human mammalian embryo, comprising
   a. Transferring the nucleus of a quiescent diploid donor cell into a suitable enucleated recipient cell of the same species, thereby obtaining a reconstituted cell;
   b. Activating the recipient cell before, during or after nuclear transfer;
   c. Incubating the reconstituted cell such that an embryo develops.

The second application, filed by the Roslin Institute, GB982916, also filed on 31 August 1995 on behalf of Ian Wilmut and Keith Campbell, describes a nuclear transfer procedure wherein a reconstituted non-human animal or embryo is produced by transferral of a diploid nucleus into an oocyte which is arrested in the metaphase of the second meiotic division without concomitantly activating the oocyte, and keeping the nucleus of the donor cell exposed to the recipient cytoplasm for a period long enough to render it capable of giving rise to a live offspring, followed by activation under conditions that maintain appropriate ploidy.

This application has been allowed in the UK, Australia and the USA. The corresponding US patent, 6,252,133 was granted with 18 claims all directed to methods for reconstructing a non-human mammalian embryo by nuclear transfer. Exemplary claim 11 of this patent is reproduced below:

11. A method of reconstructing an embryo of a non-human mammal comprising:
    a. Transferring the nucleus of a donor diploid cell in the G0 or G1 phase of the cell cycle into an unactivated, enucleated metaphase II oocyte to form a reconstructed oocyte without concomitantly activating the oocyte thereby to form a reconstructed oocyte, wherein the donor cell and the oocyte are from the same non-human mammal species;
    b. Maintaining the reconstructed embryo without activation for a period of time in the presence of at least one microtubule stabilizer or inhibitor so as to maintain correct ploidy, and wherein the resultant embryo subsequently can develop to term; and
    c. Activating and maintaining the reconstructed embryo in the presence of at least one microtubule stabilizer or inhibitor so as to maintain correct ploidy.

University of Massachusetts patent applications

The University of Massachusetts has also filed a number of patent applications (all exclusively licensed to ACT), four of which are now patented in the USA that relate to nuclear transfer procedures using non-embryonic differentiated donor cells. Particularly, the University of Massachusetts has obtained US Patents 5,945,577, 6,235,970, 6,235,969 and 6,215,041, by Cibelli *et al.*, each of which are exclusively licensed to ACT, which are directed to methods of producing cloned embryos by nuclear transfer using differentiated (somatic) donor cells or nuclei as the donor cell.

The '577 patent, issued 31 August 1999 claims a method of cloning a non-human mammal or fetus by nuclear transfer using a *proliferating* somatic cell or nucleus derived therefrom; wherein such cell has been expanded in tissue culture. Representative claim 1 is set forth below:

1. An improved method of cloning a non-human mammal by nuclear transfer comprising the introduction of a non-human mammalian donor cell or a non-human mammalian donor cell nucleus into a non-human mammalian enucleated oocyte of the same species as the donor cell or donor cell nucleus to form a nuclear transfer (NT) unit, implantation of the NT unit into the uterus of a surrogate mother of said species, and permitting the NT unit to develop into the cloned mammal, wherein the improvement comprises using as the donor cell or donor cell nucleus a proliferating somatic cell that has been expanded in culture, or a nucleus isolated from said somatic cell.

The '969 patent, issued 22 May 2001, claims methods for producing non-human mammalian embryos by nuclear transfer which utilize a 'non-quiescent mammalian differentiated cell or nucleus derived therefrom' (see claim 24 of '969 patent). Unlike the '577 patent, the patented claims do not require that the donor cell be expanded in tissue culture (thus encompassing donor cells obtained directly from donor animals) and provide for the donor cell to be any non-quiescent (proliferating) mammalian differentiated cell.

> 24. A method of producing a non-human mammalian embryo by nuclear transfer comprising transplantation of a non-human mammalian cell or a nucleus of a non-human mammalian cell into an enucleated oocyte of the same species as the donor cell or donor cell nucleus, activation of the recipient oocyte containing the donor cell or donor cell nucleus, and incubation of the activated oocyte to produce an embryo, wherein the donor cell is a non-quiescent mammalian differentiated cell or wherein the donor nucleus is from a non-quiescent mammalian differentiated cell.

The '041 patent, issued on 10 April 2001 is similarly directed to a nuclear transfer cloning method wherein the donor cell is a non-quiescent somatic cell. Exemplary claim 1 is set forth below:

> 1. An improved method of cloning a non-human mammal by nuclear transfer comprising the introduction of a non-human mammalian donor cell or a non-human mammalian donor cell nucleus into a non-human mammalian enucleated oocyte of the same species as the donor cell or donor cell nucleus to form a nuclear transfer (NT) unit, implantation of the NT unit into the uterus of a surrogate mother of said species, and permitting the NT unit to develop into the cloned mammal, wherein the improvement comprises using as the donor cell or donor cell nucleus a non-quiescent somatic cell or a nucleus isolated from said non-quiescent somatic cell.

The '970 patent differs from the previous three University of Massachusetts patents in the fact that the claims are directed to a method for providing a mammalian cultured inner cell mass by nuclear transfer. Representative of claim 1 is set forth below:

> 1. A method for producing a mammalian cultured inner cell mass (CICM) cell line by nuclear transfer comprising the following steps:
>   a. Introducing a proliferating differentiated somatic mammalian donor cell or a proliferating differentiated somatic mammalian donor cell nucleus into an enucleated mammalian oocyte of the same species to produce a nuclear transfer unit;
>   b. Activating the resultant nuclear transfer unit;
>   c. Culturing said activated nuclear transfer unit until at least a size suitable for obtaining ICM cells;
>   d. Isolating and culturing ICM cells obtained from said cultured nuclear transfer unit to obtain a cultured inner cell mass (CICM) or CICM cell line.

Cultured inner cell mass cells are disclosed in the patent to be pluripotent and therefore be useful for producing any type of cell when cultured under appropriate conditions. This is believed to be the only US patent issued to date with a claim that encompasses producing a human pluripotent cell by nuclear transfer.

Patents relating to nuclear transfer procedures by combination of donor and recipient cells of different species

The patents and applications discussed thus far all relate to nuclear transfer procedures wherein the donor cell or nucleus and the recipient cell (enucleated oocyte) are derived from the same species.

Recently, a small number of patent applications have been filed that claim nuclear transfer methods that utilize donor cells and recipient cells that are of different species origin. In particular, the University of Massachusetts has filed a number of patent applications which name Cibelli *et al.* as the inventors directed to nuclear transfer methods for producing embryonic or stem-like cells by nuclear transfer by the combination of a donor cell or nucleus of one mammalian species and an oocyte of a different animal species. An exemplary claim from one of these applications is reproduced below:

1. A method of producing embryonic or stem-like cells comprising the following steps:
   a. Inserting a desired human or mammalian cell or cell nucleus into an enucleated animal oocyte, wherein such oocyte is derived from a different animal species than the human or mammalian cell under conditions suitable for formation of a nuclear transfer (NT) unit;
   b. Activating the resultant nuclear transfer units;
   c. Culturing said activated nuclear transfer units until greater than the 2-cell developmental stage; and
   d. Culturing cells obtained from said cultured NT units to obtain embryonic or stem-like cells.

These applications exemplify the production of a human stem cell by combination of a human cell and a bovine oocyte. Some of these applications further provide for the introduction of mitochondrial DNA from the donor cell into the recipient cell to enhance cloning efficiency, as well as the introduction of heterologous genes that inhibit apoptosis and promote embryonic development. This application of NT cloning is potentially useful in situations wherein recipient oocytes are not freely available such as in the case of endangered animals or primates.

Recently, Hwang *et al.* also filed a number of patent applications in Korea, including KR 2000206 and KR 20004382 that respectively claim methods for producing cloned tigers and other endangered animals by inter-species nuclear transfer and cloned human embryos by inter-species nuclear transfer. As with the University of Massachusetts patent applications, the Hwang *et al.* application, which relates to the production of human embryos by cross-species nuclear transfer, exemplifies the production of such human embryos by combination of a human donor cell (skin cell) and an enucleated bovine oocyte (that of a Korean cow).

Infigen patent

Infigen Inc. filed a patent application in 1998, now US Patent 6,011,197, containing claims directed to cloning a bovine by nuclear transfer which requires that the donor cell is derived from a bovine fetus ranging in age from 35 to 70 days gestation and further require that the donor cell is cultured in a culture medium containing leukaemia inhibitory factor (LIF) and fibroblast growth factor (FGF) in order to effect 'reprogramming' prior to nuclear transfer. Exemplary claim 1 of this patent is set forth below:

1. A method of cloning a bovine, said method comprising:
   a. Obtaining a cell from a bovine fetus of not less than about 35 days of gestation and not more than about 70 days of gestation;
   b. Reprogramming said cell by incubating said cell in a medium comprising LIF and FGF to form a reprogrammed cell;
   c. Forming a hybrid by nuclear transfer of said reprogrammed cell into an enucleated bovine oocyte; and
   d. Culturing said cybrid so as to generate an embryo comprising embryonic cells; and
   e. Transferring said embryo of step (d) or a recloned embryo of said embryo of step (d) into the uterus of a host bovine so as to produce a fetus which undergoes full fetal development and parturition; whereby a viable clone of said bovine is generated.

Published patent applications direct to miscellaneous improvements in nuclear transfer procedures

Numerous patent applications have been filed that are directed to specific modifications and improvements in nuclear transfer procedures. Some of these patent applications are discussed below.

PPL Therapeutics filed a patent application WO 200042174, which names Keith Campbell as the inventor, that claims nuclear transfer procedures comprising the transferral of a donor nucleus into a first oocyte, followed by removal of the donor nucleus, and transferral of the nucleus into a second oocyte or an enucleated fertilized zygote. This nuclear transfer protocol is disclosed to enhance cloning efficiency, especially in the context of producing cloned transgenic animals such as pigs. The patent application indicates that their methods result in the production of non-mosaic animals and ensure the production of animals containing multiple genetic modifications.

Genetica Inc.

Genetica Inc. recently filed WO 20031238 which is directed to a method for enhancing the proliferation capacity of cells and cell lines (including cells and cell lines that are to be used as nuclear transfer donors) by contacting the cells with a first agent that reversibly activates telomerase activity and a second agent that reversibly inactivates a ras- or Rb-dependent pathway.

This procedure is reported to produce donor cells having an improved proliferative life span.

University of Edinburgh

The University of Edinburgh recently filed WO 20015473 which describes an improved method of producing transgenic animals by nuclear transfer, wherein the improvement resides in the use of donor cells which contain multiple copies of a desired heterologous DNA (this is effected by given amplification, e.g. by use of an appropriate dominant selectable marker gene). This reportedly results in cloned transgenic animals that express enhanced amounts of the polypeptide encoded by the amplified heterologous DNA.

Korea Research Institute

Korea Research Institute of Bioscience and Biotechnology recently filed WO 200135734 which is directed to an improved method of producing cloned embryos by nuclear transfer, wherein the improvement resides in the use of a first electrical simulation during fusion of the donor somatic cell and the recipient cell followed by a secondary electrical stimulation after fusion. This activation protocol is disclosed to enhance blastocyst development.

University of Georgia

The University of Georgia filed WO 2001US8371 naming Steven Stice *et al.* as the inventors that is directed to a variety of nuclear transfer protocols used to produce non-human mammalian embryos selected from the following:

1. introducing donor genetic material into a metaphase I oocyte;
2. introducing donor genetic material into a non-enucleated oocyte;
3. introducing donor genetic material into an oocyte and naturally activating the oocyte or NT embryo; or
4. introducing donor genetic material from a donor cell that is in metaphase or late G1 phase into an oocyte.

These protocols are disclosed to facilitate the exposure of the donor genetic material to the environmental factors required to reprogramme the donor nucleus and thereby enhance cloning efficiency.

University of Massachusetts

The University of Massachusetts filed WO 200126454 naming Jerry *et al.* as the inventors that claims methods for selecting cells that are in G1 cell cycle for use as donor cells in nuclear transfer procedures. The disclosed advantage is that rapidly dividing somatic cells provide for enhanced cloning efficiency.

ACT

ACT recently filed WO 200118236 directed to a nuclear transfer procedure that produces cells having enhanced life span and lengthened telomeres by the use of recloning procedures that include the use of a donor cell or nucleus that is derived from a cell cultured to near-senescence. The disclosed advantage of this method includes the fact that cells having a longer proliferative life span will be maintained for longer duration *in vivo* when used for cell therapies, and will give rise to animals having increased life span relative to the donor animal.

Stem Cell Science Party Limited

Stem Cell Science Party Limited filed WO 200067568, naming Brandon *et al.* as the inventors, is directed to a method for producing a reprogrammed diploid cell by nuclear transfer. The method comprises:

a. introducing a cell or nucleus into a recipient cell to produce an aneuploid cell and maintaining the aneuploid cell in an environment that allows the donor nucleus to be reprogrammed; and optionally activating the aneuploid cell; and
b. generating a reprogrammed diploid cell from the resultant aneuploid cell by removal, destruction or loss of the recipient cell nucleus.

Conclusions

Over the past several years there have been a number of significant innovations in the area of nuclear transfer. These innovations include in particular the use of somatic donor cells to clone non-human animals, the use of nuclear transfer to produce stable stem cell lines, the production of transgenic cloned animals by nuclear transfer, as well as various improvements in activation and fusion procedures and other techniques used in nuclear transfer procedures. These improvements and innovations are the subject of numerous pending patent applications and issued patents, some of which are discussed above. It is anticipated that the number of such patent applications and patents will only increase in number as nuclear transfer cloning becomes the established technique for producing transgenic animals of a desired genotype.

References

Baguisi, A., Behboodi, E., Melican, D.T., Pollock, J.S., Destrempes, M.M., Cammuso, C., Williams, J.L., Nims, S.D., Porter, C.A., Midura, P., *et al.* (1999) Production of goats by somatic cell nuclear transfer. *Nature Biotechnology* 17, 456–461.

Betthauser, J., Forsberg, E., Augenstein, M., Childs, L., Eilertsen, K., Enos, J., Forsythe, T., Golueke, P., Jurgella, G., Koppang, R., *et al.* (2000) Production of cloned pigs from in vitro systems [In Process Citation]. *Nature Biotechnology* 18, 1055–1059.

Bondioli, K., Ramsoondar, J., Williams, B., Costa, C. and Fodor, W. (2001) Cloned pigs generated from cultured skin fibroblasts derived from a H-transferase transgenic boar. *Molecular Reproduction and Development* 60, 189–195.

Briggs, R. and King, T. (1952) Transplantation of living nuclei from blastula cells into enucleated frogs' eggs. *Proceedings of the National Academy of Sciences USA* 38, 455–463.

Campbell, K.H.S., Ritchie, W.A. and Wilmut, I. (1993) Nuclear-cytoplasmic interactions during the first cell cycle of nuclear transfer reconstructed bovine embryos: implications for deoxyribonucleic acid replication and development. *Biology of Reproduction* 49, 933–942.

Cibelli, J.B., Stice, S.L., Golueke, P.J., Kane, J.J., Jerry, J., Blackwell, C., Ponce de Leon, F.A. and Robl, J.M. (1998) Cloned transgenic calves produced from nonquiescent fetal fibroblasts. *Science* 280,1256–1258.

Denning, C., Burl, S., Ainslie, A., Bracken, J., Dinnyes, A., Fletcher, J., King, T., Ritchie, M., Ritchie, W.A., Rollo, M., *et al.* (2001) Deletion of the alpha(1,3)galactosyl transferase (GGTA1) gene and the prion protein (PrP) gene in sheep. *Nature Biotechnology* 19, 559–562.

Galli, C., Duchi, R., Moor, R. and Lazzari, G. (1999) Mammalian leukocytes contain all the genetic information necessary for the development of a new individual. *Cloning* 1, 161–170.

Gurdon, J.B., Laskey, R.A. and Reeves, O.R. (1975) The development capacity of nuclei transplanted from keratinized skin cells of adult frogs. *Journal of Embryology and Experimental Morphology* 34, 93–112.

Hill, J.R., Winger, Q.A., Long, C.R., Looney, C.R., Thompson, J.A. and Westhusin, M.E. (2000) Development rates of male bovine nuclear transfer embryos derived from adult and fetal cells. *Biology of Reproduction* 62, 1135–1140.

Hoppe, P.C. and Illmensee, K. (1982) Full-term development after transplantation of parthenogenetic embryonic nuclei into fertilized mouse eggs. *Proceedings of the National Academy of Sciences USA* 79, 1912–1916.

Kasinathan, P., Knott, J.G., Wang, Z., Jerry, D.J. and Robl, J.M. (2001) Production of calves from G1 fibroblasts. *Nature Biotechnology* 19, 1176–1178.

Keefer, C.L., Baldassarre, H., Keyston, R., Wang, B., Bhatia, B., Bilodeau, A.S., Zhou, J.F., Leduc, M., Downey, B.R., Lazaris, A., *et al.* (2001) Generation of dwarf goat (*Capra hircus*) clones following nuclear transfer with transfected and nontransfected fetal fibroblasts and *in vitro*-matured oocytes. *Biology of Reproduction* 64, 849–856.

Lanza, R.P., Cibelli, J.B., Faber, D., Sweeney, R.W., Henderson, B., Nevala, W., West, M.D. and Wettstein, P.J. (2001) Cloned cattle can be healthy and normal. *Science* 294, 1893–1894.

Liu, L., Dai, Y. and Moor, R.M. (1997) Nuclear transfer in sheep embryos: the effect of cell cycle coordination between nucleus and cytoplasm and the use of *in vitro* matured oocytes. *Molecular Reproduction and Development* 47, 255–264.

Loi, P., Ptak, G., Barboni, B., Fulka, J., Jr, Cappai, P. and Clinton, M. (2001) Genetic rescue of an endangered mammal by cross-species nuclear transfer using post-mortem somatic cells. *Nature Biotechnology* 19, 962–964.

McCreath, K.J., Howcroft, J., Campbell, K.H., Colman, A., Schnieke, A.E. and Kind, A.J. (2000) Production of gene-targeted sheep by nuclear transfer from cultured somatic cells. *Nature* 405,1066–1069.

McGrath, J. and Solter, D. (1983) Nuclear transplantation in mouse embryos. *Journal of Experimental Zoology* 228, 355–362.

Ogura, A., Inoue, K., Takano, K., Wakayama, T. and Yanagimachi, R. (2000a) Birth of mice after nuclear transfer by electrofusion using tail tip cells. *Molecular Reproduction and Development* 57, 55–59.

Ogura, A., Inoue, K., Ogonuki, N., Noguchi, A., Takano, K., Nagano, R., Suzuki, O., Lee, J., Ishino, F. and Matsuda, J. (2000b) Production of male cloned mice from fresh, cultured, and cryopreserved immature Sertoli cells. *Biology of Reproduction* 62, 1579–1584.

Onishi, A., Iwamoto, M., Akita, T., Mikawa, S., Takeda, K., Awata, T., Hanada, H. and Perry, A.C. (2000) Pig cloning by microinjection of fetal fibroblast nuclei. *Science* 289,1188–1190.

Ono, Y., Shimozawa, N., Ito, M. and Kono, T. (2001) Cloned mice from fetal fibroblast cells arrested at metaphase by a serial nuclear transfer. *Biology of Reproduction* 64, 44–50.

Polejaeva, I.A., Chen, S.H., Vaught, T.D., Page, R.L., Mullins, J., Ball, S., Dai, Y., Boone, J., Walker, S., Ayares, D.L., *et al.* (2000) Cloned pigs produced by nuclear transfer from adult somatic cells. *Nature* 407, 86–90.

Reggio, B.C., James, A.N., Green, H.L., Gavin, W.G., Behboodi, E., Echelard, Y. and Godke, R.A. (2001) Cloned transgenic offspring resulting from somatic cell nuclear transfer in the goat: oocytes derived from both follicle-stimulating hormone-stimulated and nonstimulated abattoir-derived ovaries. *Biology of Reproduction* 65, 1528–1533.

Renard, J.P., Chastant, S., Chesne, P., Richard, C., Marchal, J., Cordonnier, N., Chavatte, P. and Vignon, X. (1999) Lymphoid hypoplasia and somatic cloning. *Lancet* 353,1489–1491.

Schnieke, A.E., Ritchie, W.A., Mycock, K., Scott, A.R., Ritchie, M., Wilmut, I., Colman, A. and Campbell, K.H.S. (1997) Human factor IX transgenic sheep produced by transfer of nuclei from transfected fetal fibroblasts. *Science* 278, 2130–2133.

Shiga, K., Fujita, T., Hirose, K., Sasae, Y. and Nagai, T. (1999) Production of calves by transfer of nuclei from cultured somatic cells obtained from Japanese black bulls. *Theriogenology* 52, 527–535.

Spemann, H. (1914) *Ver. Deutsch. Zool. Ges. Leipzig (Freiburg)* 24, 216–221.

Spemann, H. (1938) *Embryonic Development and Induction.*

Tamashiro, K.L., Wakayama, T., Blanchard, R.J., Blanchard, D.C. and Yanagimachi, R. (2000) Postnatal growth and behavioural development of mice cloned from adult cumulus cells. *Biology of Reproduction* 63, 328–334.

Wakayama, T. and Yanagimachi, R. (1999) Cloning of male mice from adult tail-tip cells. *Nature Genetics* 22, 127–128.

Wakayama, T. and Yanagimachi, R. (2001) Mouse cloning with nucleus donor cells of different age and type. *Molecular Reproduction and Development* 58, 376–383.

Wakayama, T., Perry, A.C., Zuccotti, M., Johnson, K.R. and Yanagimachi, R. (1998) Full-term development of mice from enucleated oocytes injected with cumulus cell nuclei. *Nature* 394, 369–374.

Willadsen, S.M. (1986) Nuclear transplantation in sheep embryos. *Nature* 320, 63–65.

Wilmut, I., Schnieke, A.E., McWhir, J., Kind, A.J. and Campbell, K.H.S. (1997) Viable offspring derived from fetal and adult mammalian cells. *Nature* 385, 810–813.

Yamazaki, Y., Makino, H., Hamaguchi-Hamada, K., Hamada, S., Sugino, H., Kawase, E., Miyata, T., Ogawa, M., Yanagimachi, R. and Yagi, T. (2001) Assessment of the developmental totipotency of neural cells in the cerebral cortex of mouse embryo by nuclear transfer. *Proceedings of the National Academy of Sciences USA* 98, 14022–14026.

Zakhartchenko, V., Alberio, R., Stojkovic, M., Prelle, K., Schernthaner, W., Stojkovic, P., Wenigerkind, H., Wanke, R., Duchler, M., Steinborn, R., *et al.* (1999) Adult cloning in cattle: potential of nuclei from a permanent cell line and from primary cultures. *Molecular Reproduction and Development* 54, 264–272.

Zakhartchenko, V., Mueller, S., Alberio, R., Schernthaner, W., Stojkovic, M., Wenigerkind, H., Wanke, R., Lassnig, C., Mueller, M., Wolf, E., *et al.* (2001) Nuclear transfer in cattle with non-transfected and transfected fetal or cloned transgenic fetal and postnatal fibroblasts. *Molecular Reproduction and Development* 60, 362–369.

Zhou, Q., Boulanger, L. and Renard, J.-P. (2000) A simplified method for the reconstruction of fully competent mouse zygotes from adult somatic donor nuclei. *Cloning* 2, 35–44.

Zou, X., Wang, Y., Cheng, Y., Yang, Y., Ju, H., Tang, H., Shen, Y., Mu, Z., Xu, S. and Du, M. (2002) Generation of cloned goats (*Capra hircus*) from transfected foetal fibroblast cells, the effect of donor cell cycle. *Molecular Reproduction and Development* 61, 164–172.

14

Dairy Cattle Test Day Models: a Case Study

Lawrence R. Schaeffer

Centre for Genetic Improvement of Livestock, Department of Animal and Poultry Science, University of Guelph, Guelph, Ontario NIG 2WI, Canada

Introduction

This case study is an example of a patent that has met with considerable resistance amongst the animal breeding research community and its validity has been challenged in Canada and the USA. Dr Robert W. Everett of Cornell University and the Cornell Research Foundation, Inc. filed for a patent entitled 'Method of Bovine Herd Management' on 25 February 1993 in the USA, with the Canadian application filed on 15 February 1994. The patents were granted on 4 October 1994 and 14 July 1998 for the USA and Canada, respectively. The Cornell Research Foundation is on the verge of obtaining a successful application in the European Union. The details of the patent will be discussed. Background information on the dairy cattle industry and milk recording programmes is needed to understand the claims in the patents and to judge its validity.

Background

The beginning of milk testing

In dairy cattle, herds enrolled on a milk recording programme are visited approximately once per month by a field supervisor. These visits have been called test days (TD). The supervisor weighs the milk for each cow at each milking, takes a sample for fat and protein analyses, and records other events such as calvings, dry off dates and culling information. TD information is processed, validated and stored by the milk recording centres to produce a summary of the amount of milk given by each cow to date, and a projected 305-day lactation total yield per cow and per herd. Herd owners utilize these statistics for management purposes to adjust feeding programmes, to plan milk income and to cull poor-producing cows from the herd. Dairy producers pay milk-recording organizations for these services. This routine has been commonplace in milk recording programmes around the world since the early 1900s. Evidence of this early history comes from the *National Cooperative Dairy Herd Improvement Program Hand Book*, which contains a history of dairy record keeping written by D.E. Voelker. The following is a quote from that section (kindly provided by A.E. Freeman, 14 July 2000):

> In September 1905, a Danish immigrant, Helmer Rabild, working as an inspector for the Michigan Dairy and Food Commission, called a meeting and six to eight dairy farmers from Newaygo County, Michigan attended. As the result of Rabild's knowledge and enthusiasm, the first cow testing association was formed. In 1906, the association's members represented 31 herds and 239 cows. In 1908, when Rabild was working for the dairy division of the U.S. Department of Agriculture (USDA), similar cow testing associations were started in Maine and New York.

There is no mention of TD records, but Voelker stated that the standard lactation length of 305-days was established in 1935. Most likely, Denmark or other European countries must have had cow-testing associations prior to 1905.

Genetic evaluation methods using 305-day yields

Geneticists began to utilize the 305-day lactation yields of cows obtained from milk recording organizations in order to evaluate the genetic merit of dairy bulls almost as soon as the data started to accumulate around 1935 in the USA. The genetic evaluations were important tools for the artificial insemination (AI) industry to select bulls with superior genetics for production traits. The methodology to analyse the 305-day lactation records has evolved over time to become more and more sophisticated. The application of new genetic evaluation methodology has closely followed the evolution of computing hardware. New technological advances have allowed the application of new theories of genetic evaluation that were previously considered impractical or even impossible. The key issue here is that ideas about statistical models for genetic evaluation purposes generally preceded the applications of those models to real data due to lack of adequate computing hardware to do the job efficiently.

Initial genetic evaluations were daughter–dam comparisons based on simple averages of differences in daughter and dams records (Lush, 1931, 1933). Eventually, herd management was found to be a major influence on daughter–dam comparisons. The dam's records were made several years ahead of those of their daughters. The use of a contemporary herd average was proposed by Henderson *et al.* (1954). The Herdmate Comparison procedure was adopted in the USA in 1962 and compared cows within herd–year–season of calving groups, and combined these deviations across herds for bulls' daughters. The Herdmate Comparison did not account for the merit of a bull's mates, the genetic level of competition within a herd–year–season, or for differential culling among daughters of bulls. The Modified Contemporary Comparison method followed in 1974 that was a patchwork of corrections to the Herdmate Comparison.

A few years earlier (1971), Dr C.R. Henderson of Cornell University, working with Eastern AI Cooperative in Ithaca, New York, developed a procedure based on best linear unbiased prediction or BLUP methodology (Henderson, 1973). BLUP methods were actually first known to econometrists in 1962 (Goldberger, 1962), but Henderson is generally credited with promoting their use in animal breeding since 1963. The advantage of BLUP was that it forced the user to articulate a statistical model that described how the 305-day lactation records were sampled and the factors that could influence yields. The model in 1971 was called a Sire Model. Sires were assumed to be unrelated to each other, and to be randomly mated to dams. Each dam was assumed to have only one progeny. The results were published and called the Northeast AI Sire Comparison by the Northeast Dairy Records Processing Laboratory (DRPL) located at Cornell University. Henderson (1975, 1976) discovered a computational shortcut that allowed genetic relationships among animals to be included in BLUP methodology, and therefore, relationships among sires could be part of the genetic analyses. This discovery was implemented into the Northeast AI Sire Comparison immediately. The success of these models on the genetic progress in the northeastern USA was quantified by Everett and Keown (1984).

Although Henderson first described an Animal Model around 1976, it was not until 1985 that Animal Models were applied to national dairy milk records in Australia, and later in 1989 in Canada and the USA, due to the need for computer hardware to catch up to statistical and genetic theory. Animal Models seemingly solved many problems of bias caused by non-random mating. *Interbull Bulletin No. 5* (1992) summarized the genetic evaluation procedures used by over 28 countries in the world, and 14 of these utilized an Animal Model of one variation or another with BLUP methodology. The latest *Interbull Bulletin No. 24* (2000) shows that nearly all countries use an Animal Model today, except for a few which have progressed to using test day records.

From 1935 to 1999, genetic evaluations in dairy cattle have taken 305-day lactation yields as the standard observation for analysis. A 305-day lactation yield of a cow, however, is estimated from the monthly test day information collected during her lactation. There could be from 7 to 10 or more test day records per cow that are combined to provide an estimate of the total amount of milk given in 305 days. At one time in Canada, there was a recording programme which collected daily milk yields of cows in particular herds. The 305-day yields were simply the sum of those daily yields. Because the collection of daily yields was very expensive, Canada and the majority of countries collect just monthly test day records.

Making use of test day records

One of the major contributors to milk recording programmes in Canada was Dr John Moxley of McGill University in Montreal. In 1968, he presented a paper entitled 'Within herd evaluation in the Dairy Herd Analysis Service Program' to the Canadian Society of Animal Production Meeting. The abstract of that paper was as follows:

> The Dairy Herd Analysis Service provides a cow rating for each cow in the herd on the monthly test day report. This cow rating is based on 4% fat corrected milk and the rating takes into consideration age, stage of lactation and seasonal effects peculiar to the herd. The rating for each cow is first reported after the cow has had normal tests and is updated on each subsequent test day. Genetic changes in the herd are automatically adjusted for on each test day by using 100 as herd average on each test day. The DHAS cow rating provides a basis for culling while the lactation is in progress. The monthly-adjusted herd average test day yield is useful in evaluating seasonal changes in herd feeding and management. Complete and part lactations can be appropriately weighted to provide efficient ratings of AI sires.

Clearly, Moxley was promoting the use of TD information, and adjusting those yields for age and stage of lactation. In meetings of the Canadian Dairy Technical Committee (which reported to Agriculture Canada at that time), Moxley suggested that AI sires could be evaluated using TD records directly (mostly to avoid the confusion about extending part lactation records to a 305-day basis). However, computing technology in 1968 was not capable of handling the very large number of TD records, roughly 10 times the number of 305-day records. Also, this was before Henderson's BLUP methodology was widely known and before appropriate methods for analysing TD records were known. The suggestion was not feasible at that time.

P.D.P. Wood (1967, 1968) began publishing his work with lactation curves making use of TD records of dairy cows. The idea was to find a way of predicting the total 305-day yield from a few TD records. Differences in the shapes of curves between cows were noted, and the possibility of selecting for the shape of the lactation curve was put in front of researchers. Wood also noted that the diet of the cow could affect the shape of the curve. All of this work was done on a phenotypic basis rather than genetic, and either on individual cows or groups of cows. Many other functions have been applied to this problem since 1967 (Schaeffer and Burnside, 1976; Schaeffer *et al.*, 1977; Ali and Schaeffer, 1987).

When did TD records start to be used for herd management or genetic evaluation? The following message was sent to the Animal Geneticists Discussion Group on 11 July, 2000 from Les Jones of the Australian Dairy Herd Improvement Service.

> In Australia, farmers have been using production indexes at least since the early 1980s. In this index, the test day yield of a cow on a particular day is corrected for age and stage of lactation and then compared with the average corrected yield on the particular day. The deviations are then combined into an overall measure of merit called a Production Index. Farmers make use of this information in making culling decisions. This approach has had considerable benefit under Australian grazing conditions, where average yields between test days can differ markedly.
>
> My colleague, Kevin Beard, has shown me a covering letter dated 3 August 1981 to a herd recording organization providing information on the Production Index. Kevin examined the appropriate correction factors in his Master's thesis at Melbourne University, submitted in 1983.
>
> The Australian Dairy Herd Improvement Scheme has been using test days in a similar way for the calculation of breeding values since 1985, a year after we started using the Animal Model. (Proceedings of 5th Conference of Australian Association of Animal Breeding and Genetics, 1985. Jones, L. P. Australian breeding values for production characters. pp. 242–247)
>
> Credit for the idea should go to my predecessor, Geoff Robinson. He had planned its use for this purpose before I took over the position in 1982. Clearly this approach makes less use of the test day records than the more recent test day models. It enables us to use a longer herd-year-season than previously, but does not compute estimates of persistency.

See also Jones and Goddard (1990).

TD models

After 1989, computer hardware rapidly changed. Processing speeds increased, computer random access memory changed from kilobytes to megabytes to gigabytes, and disk capacity was skyrocketing. Computing moved from mainframes in big air-conditioned rooms to workstations on researchers' desks. By 1991, Moxley's 1968 proposal to directly use TD records in genetic evaluation of AI sires started to fall within the realm of possibility. For anyone deeply involved in milk recording, the idea of analysing TD records would have been a natural and obvious progression. Also, dairymen started to think more about the costs associated with monthly testing and were asking if supervised tests were essential for accurate estimation of 305-day yields. Could recording schemes be modified to have a combination of supervised and unsupervised recording? Could the interval between TDs be greater than 30 days and could the number of TD records per lactation be decreased? Do fat and protein content have to be determined at each test? Kachman and Everett (1989) showed how to correct TD records per cow within herds, followed by work on lactation curves and heritabilities of TD yields (Stanton *et al.*, 1992), and coined the term Test Day Models (TDMs). Ptak and Schaeffer (1993) presented a TDM for Canadian dairy cattle based on work started in 1991. The shift to TDM analyses based on the recognized growth in computer hardware developments, and based on a need to reduce the costs of milk recording for dairy producers was an inevitable and obvious occurrence as evidenced by early studies in Canada, Australia and the USA.

Connectedness of researchers

An important question was whether the developments in Canada, Australia and the USA were independent of each other. The actual details of how each country proceeded to TD record analyses were very different in terms of models and procedures, but the idea to work with TD records may not have been independent. The interrelationships between scientists involved in this research were very interwoven. Moxley and Schaeffer were students of Henderson at Cornell, although at different periods, and Everett was a faculty colleague at Cornell. As a graduate student, Schaeffer worked with Everett on TD records for studies on days open. In Canada, Schaeffer and Moxley participated together on the Canadian Dairy Technical Committee in the mid 1970s. Discussion of Moxley's proposal to use TD records arose more than once during those semi-annual meetings. Everett was deeply involved in the DRPL lab and milk recording issues in the northeastern USA. Interestingly, Everett took two sabbatical leaves in Australia, first in 1980–1981 and later in 1987–1988, and had to be aware of the Australian use of TD records and their Production Index during those visits.

The University of Guelph and Cornell University held annual colloquia through the 1980s and 1990s. Thus, the research efforts of both groups were known to each other. Whether or not Moxley had any influence on Everett or the Australians is not known, but being deeply involved in milk recording programmes in their respective institutions they were probably aware of each others' programmes for dairy producers. In my mind, the credit for using TD records in genetic evaluation should ultimately be given to Moxley, but exactly who was first to think about this is not absolutely clear.

The Canadian TDM

Canada introduced a multiple trait, random regression TDM in February 1999 after 8 years of research and trial runs. This system utilizes TD records from the first three lactations between 5 and 305 days in milk. Milk, fat, protein yields and somatic cell scores are analysed as separate traits within each lactation. Thus, there are 12 traits in total, assuming that the shapes of their lactation curves were different. A three-parameter linear function is used to define the general shape of the curves within time periods, regions of Canada, parities, seasons of calving and age at calving. Random regressions are in the model to account for the genetic and environmental cow-to-cow variation around the standard curves. Research on this model is continuing and many improvements can be made. Details of Canada's experience with this model can be found in Schaeffer *et al.* (2000). The Canadian TDM fulfilled some of the concerns of dairy producers in that cows do not need to have monthly TD records per lactation, and not all TD need to measure fat and protein content. Four TD records in the TDM seem to supply as much information as 10 TD records in the previous 305-day lactation Animal Model system. Several countries have already adopted a TDM or are in the research stages towards a TDM.

Other countries (the USA, Germany, Finland, the Netherlands, Italy and the UK) have also been developing TDM during the same time, but have not yet officially adopted a TDM. In the USA, the Everett–Cornell patent has stalled implementation of a TDM. Outside of the USA, the attitude of researchers and organizations that compute genetic evaluations of dairy cattle has been defiant in the belief that the patent was not valid and that eventually this belief would be proved true.

Summary of background

TD records have been collected by milk recording agencies around the world for at least 100 years. Since 1935, TD records have been combined into standard 305-day lactation yields that have been used to evaluate dairy bulls genetically so that dairymen could make informed mating decisions about their cows. The manner in which genetic evaluations have been computed has evolved over the years as the result of mutual exchanges and publication of research information that has occurred at a number of institutions around the world, and as a result of technological advances in computer hardware and software. TD records may have first been utilized in Australia in the 1980s for herd management and in 1985 for genetic evaluations. The latest genetic evaluation procedures, based on Moxley's 1968 proposal, have been directly to analyse TD records in what are generally termed TDMs, of which the Canadian TDM is just one example.

The Everett–Cornell Patent

Given the previous background information about milk recording and genetic evaluation systems, one might wonder what aspects could be patentable, who would want to patent anything related to a system that has been viable for almost 100 years, and why? When one thinks about a patent, there is the notion that there must be something new that has been invented that did not exist previously. An invention is usually thought to be a physical entity like a light bulb or mousetrap. The Everett–Cornell patent does not seem to fit the conventional notions about patents. The invention in this case is a 'Method of Bovine Herd Management' which includes the gathering of data, the mathematical treatment of that data, and the subsequent use of the modified data by dairy producers. Thus, nothing physical has been created or built, but rather the invention is a system or process. A process is patentable if it is novel and not obvious. Are those criteria met in this case?

Gathering of data

The Method of Bovine Herd Management includes the gathering of TD data on a routine basis. This includes all quantitative and qualitative milk production data for each member of a herd. Qualitative milk production data refers to information about number of times the cow was milked on that day, her pregnancy status, health status, temperament, milking speed, birth date, sire and dam identification, breed composition, and any other information that may become important in the future. Quantitative data refers to the amount of milk, the percentages of fat, protein, lactose and other constituents of milk, as well as somatic cell counts. Milk recording organizations have been gathering data in this manner since the early 1900s.

The collection of data requires the establishment and continuous updating of a database for each cow within a herd. Databases from several herds could be linked or combined for the purpose of computing genetic evaluations. Conceptually, the database could be national in scope as well as the genetic evaluation system. These elements are also included in the patent. Note that Moxley described this exactly in his 1968 presentation to the Canadian Society of Animal Production. Thus, the gathering of data cannot be considered novel.

Modifying TD data

The Method of Bovine Herd Management includes the use of a mathematical herd management model to modify TD data to determine the actual productivity of each cow in the herd. The patent claims that up to 40% of the environmental variation in TD yield can be removed. The model, as given in the patent for the purpose of disclosure only, is

$$y_{ijklmn} = H_i + A_j + D_k + P_l + F_m + e_{ijklmn}$$

where y_{ijklmn} is the TD observation on a production trait (milk, fat or protein yields, or somatic cell scores), H_i is a TD effect within herd (identifies all of the contemporaries in the herd on TD as well as the environment), A_j is the effect of age (in months) on the day observed, D_k is an effect of days in milk when observed, P_l is an effect of number of days pregnant when observed, F_m is an effect of month of freshening, and e_{ijklmn} is a random residual effect.

Although not stated in the patent explicitly, only first lactation cows are intended to be included in these analyses. Otherwise, the model would need to have lactation effects and interactions of these with age at calving and days in milk because cows in different lactations, calving at different ages would have significantly different shapes of lactation curves. The residual effects are correlated within a cow, and the patent implies that an autocorrelation structure is assumed between TDs on the same cow.

After estimating the differences in effects in the model by generalized least squares, then residuals, $\hat{e}_{ijklmn}$, are estimated as

$$\hat{e}_{ijklmn} = (y_{ijklmn} - H_i - A_j - D_k - P_l - F_m).$$

Finally, the residuals are combined into a total lactation value, $\hat{t}_n$, based on n TD records, as

$$\hat{t}_n = 1'RM(M'RM)^{-1}\hat{e}$$

where $\hat{e}$ is a vector of residuals for a cow, R is the autocorrelation structure of the residuals, and M is described as a mapping incidence matrix, but is not explicitly given in the example. M must contain covariances between the observed TD records and total 305-day yields. The modified total yields, $\hat{t}_n$, are also stored in the database along with the original TD yields.

The patent discusses primarily the application to milk yields, but the same procedure is assumed to be applied separately to fat and protein yields, and possibly other traits. A further complexity to this approach would be a multiple trait application to all production traits simultaneously. Also, the patent discusses only the use on dairy cows, but TD models have been applied to dairy goats and dairy sheep too. Whether or not the patent covers these species is not clear. Finally, could this patent also cover any longitudinal data collected on cows, such as body weights, condition scores, feed intakes, milk urea nitrogen levels, and so on?

The specific within herd model, as described in the patent, was novel at the time. All other developments concerning TD models that have been reported in the literature before and after the patent would not be in conflict with the patent because these models are quite different from that in the patent description. However, there is an 'include all' statement in the patent that reads as follows:

> Since other modifications and changes varied to fit particular operating requirements and environments will be apparent to those skilled in the art, the invention is not considered limited to the example chosen for purposes of disclosure, and covers all changes and modifications which do not constitute departures from the true spirit and scope of this invention.

Therefore, any kind of TDM applied to TD data on dairy cows would be an infringement of the patent rights. This is analogous to someone inventing a mousetrap of one particular design, but the patent also covers any other future, possibly different, design of mousetraps whose true spirit is to catch mice. How can a patent claim rights to all future, obvious developments to improve upon the model that was disclosed?

The word 'art' should be replaced by 'science'. Through science, models are developed based on statistical tests and comparisons for the better fit to the data. Hopefully new models explain the biology of the situation too. Should it be permitted to patent a statistical model? Statistical models always have a degree of error in explaining the variation in a set of data. If a patent is granted for a particular model, then should that patent include claims for any future enhanced models for analysing the same kind of data?

Use of modified TD records

The patent describes the use of modified TD records by herd owners to make decisions that could improve the milk productivity of the herd. The decisions that could be made would be

1. To cull the less productive members from the herd;
2. To breed the more productive members to the genetically superior AI bulls;
3. To change the feed formulation and management practices of the herd;
4. To change the environmental conditions in which an animal or group of animals are housed; and
5. To mate AI bulls with cows from different herds based on genetic evaluations from the TD data.

Dairy producers have been striving to maximize milk production for over a century through the use of information obtained from a milk recording organization and from national genetic evaluation systems that utilize data from all milk recording organizations in that country. Obviously, genetic evaluations from TD models will be more accurate and will result in better-informed decisions by dairy producers, but this is just a part of the evolution of genetic evaluation models that began back in 1931. Everett did not 'invent' anything new, but has merely been one participant in the general evolution of statistical models and procedures to the analysis of TD records. Everett's model is drastically different from the Canadian TDM of February 1999, for example, but in true spirit the Canadian TDM is aimed at the accurate evaluation of genetic merit of bulls and cows using TD records based on an idea from 1968.

Dairy producers utilize more than just TD records and genetic evaluations in their decision-making process. Producers want to maximize profitability, and factors they include besides productivity are conformation, health, reproduction, temperament, milking speed, labour and feed intake. The amount of emphasis placed on milk production would undoubtedly vary from one herd to the next. Ironically, even herd owners that do not participate in milk recording activities can benefit from the genetic evaluations of bulls based on the TD data. The patent cannot really dictate how TD information would be used, but rather the patent wants to limit any organization from collecting, analysing and dispersing the results without permission from or re-imbursement of a fee to the holders of the patent.

Concluding Remarks

There is an e-mail newsletter produced by Gregory Aharonian that questions the quality of patents that are granted in the USA. He makes a living by searching the literature on the originality of patent applications. A discussion with Aharonian appears in the December 2001 issue of *Scientific American* (Millette, 2001). Aharonian claims that patents are issued without looking at the literature for prior art, and he is publicizing his results. The problems seem to be the growth rate in patent applications, the difficulty in hiring quality examiners, and the lack of funding going to the Patent and Trademark Office. Examiners do not have the time and resources to seek prior art, and applicants are refusing to do much searching on their own. Consequently, a patent is granted that covers all TDM methods and use of TD records instead of a patent that is more narrowly focused on one particular TDM method.

The novelty and obviousness of the Everett–Cornell patent has been seriously questioned. Clearly, the practices of gathering TD data, manipulating it into various forms for use by dairy producers, and the dispersal of TD information to producers have existed for nearly 100 years. The patent claims rights to a practice that has been public knowledge for a long time. The only novel idea within the patent was the specific mathematical model and procedures that Everett and co-workers developed for the analysis of TD yields. Everett was not the first researcher to apply a model to TD records. The model, as described in the patent, is not necessarily the best model that could be applied. Details have been omitted or simplified in the patent description. If the Everett model had explained 100% of the variation in TD records, then a patent would be justified, but unfortunately, every statistical model is just an approximation to reality, and has a degree of error associated with it. Most scientists publish their models in order to stimulate other researchers to find fault or to improve upon their model.

In March 2001, the Canadian Dairy Network submitted a formal request to the Canadian Intellectual Property Office for a re-examination of the validity of the Everett–Cornell patent. In accordance with the Patent Act, the Canadian Intellectual Property Office appointed a Re-examination Board in June 2001, and on 31 October 2001 this board concluded the following:

> In accordance with Section 48.2(2) of the Patent Act, the Re-examination Board has reviewed the request for re-examination of patent 2,130,471 and has determined that it raises a substantial new question of patentability regarding claims 1 through 19.

The Cornell Research Foundation is allowed to respond with its own submission and arguments prior to 31 January 2002. The final decision is expected by 31 March 2002.

Similarly, the USDA asked for a review of the validity of the patent a few years ago, but the judge upheld the patent. Clearly, legal viewpoints about the patent are different from the 'logical' viewpoints of animal breeders. From a scientific viewpoint, the claims in the patent are a copy of other people's ideas, except for the particular mathematical model (and even that may have been borrowed from the Australians).

Several countries in Europe have proceeded to implement a TDM for genetic evaluation of their cattle. Germany, Finland, the Netherlands, Italy and the UK are proceeding with their own TDMs. Approval of the Everett–Cornell patent in Europe should be much more difficult than in North America. According to Article 52 of the European Patent Office (www.european-patent-office.org/online):

1. European patents shall be granted for any inventions which are susceptible to industrial application, which are new and which involve an inventive step.
    a. The following in particular shall NOT be regarded as inventions within the meaning of paragraph 1.
    b. Discoveries, scientific theories and mathematical models; (b) Aesthetic creations;
2. Schemes, rules and methods for performing mental acts, playing games or doing business, and programmes for computers;
3. Presentation of information.

The provision of paragraph 2 shall exclude patentability of the subject-matter or activities referred to in that provision only to the extent to which a European patent application or European patent relates to such subject-matter or activities as such.

However, the European Patent Office is close to approving the Everett–Cornell patent. Why? If the patent is granted in the European Union, then there could be many legal battles with each individual country in the years to come.

If the patent was valid in every country, then there could be serious implications for genetic evaluation programmes. For Canada, royalties would have to be paid to the Cornell Research Foundation and maybe some penalty fees that Canada can ill afford. The Genetic Evaluation Board in Canada would have to decide whether to continue with a TDM and pay a licence fee to Cornell or revert back to a less accurate animal model using 305-day yields. In other countries, research on TDMs may cease altogether as no one will be able to or want to pay royalties. This would be a serious setback to the overall genetic improvement of dairy cattle around the world, except for animals evaluated by Cornell University.

The best-case scenario is that the Everett–Cornell patent will be found invalid due to the existence of published documents and scientific papers as evidence of prior art. Hopefully, if the Canadian Dairy Network is successful in its re-examination of the patent, then they will share their legal documentation with other countries. The Everett–Cornell patent is an example of a case where the examiners apparently failed to look up the literature on prior art, or if they did, then they did not understand this area of expertise or the background history behind milk recording in dairy cattle. Either way, a patent was granted that should have been thrown out upon the first reading of it.

Note Added in Proof

As of 1 July 2002, the Cornell Research Foundation has been allowed additional time to respond to the Canadian challenge and the Re-examination Board will not make a final decision until October 2002. The European Union approved the Everett–Cornell patent in June 2002. Companies within the EU have 3 months to challenge the patent.

References

Ali, T.E. and Schaeffer, L.R. (1987) Accounting for covariances among test day milk yields in dairy cows. *Canadian Journal of Animal Science* 67, 637–644.

Everett, R.W. and Keown, J.F. (1984) Mixed model sire evaluation with dairy cattle – experience and genetic gain. *Journal of Animal Science* 59, 529–541.

Goldberger, A.S. (1962) Best linear unbiased prediction in the generalized linear regression model. *Journal of the American Statistical Association* 57, 369.

Henderson, C.R. (1973) Sire evaluation and genetic trends. *Proceedings of the Animal Breeding and Genetics Symposium in Honor of Dr. Jay L. Lush*. American Society of Animal Science, Urbana, Illinois, and American Dairy Science Association, Champaign, Illinois, pp. 10–41.

Henderson, C.R. (1975) Use of relationship among sires to increase accuracy of sire evaluation. *Journal of Dairy Science* 58, 1731–1738.

Henderson, C.R. (1976) A simple method for computing the inverse of a numerator relationship matrix used in prediction of breeding values. *Biometrics* 32, 69–83.

Henderson, C.R., Carter, H.W. and Godfrey, J.T. (1954) Use of contemporary herd average in appraising progeny tests of dairy bulls. *Journal of Animal Science* 13, 949.

Interbull (1992) Sire evaluation procedures for dairy production traits practiced in various countries. *Bulletin No. 5.*

Interbull (2000) National genetic evaluation programmes for dairy production in Interbull member countries 1999–2000. *Bulletin No. 24.*

Jones, L.P. and Goddard, M.E. (1990) Five years experience with the animal model for dairy evaluations in Australia. *Proceedings of the 4th World Congress on Genet. Appl. Livest. Prod.*, Edinburgh, UK, XIII, pp. 382–385.

Kachman, S.D. and Everett, R.W. (1989) Test day model with individual herd correction factors. *Proceedings of Annual Meeting of the American Dairy Science Association, Journal of Dairy Science,* p. 60 (Abstract).

Lush, J.L. (1931) The number of daughters necessary to prove a sire. *Journal of Dairy Science* 14, 209–220.

Lush, J.L. (1933) The bull index problem in the light of modern genetics. *Journal of Dairy Science* 16, 501–522.

Millette, E. (2001) Patent pamphleteer. *Scientific American*, p. 33.

Ptak, E. and Schaeffer, L.R. (1993) Use of test day yields for genetic evaluation of dairy sires and cows. *Livestock Production Science* 34, 23–34.

Schaeffer, L.R. and Burnside, E.B. (1976) Estimating the shape of the lactation curve. *Canadian Journal of Animal Science* 56, 157–170.

Schaeffer, L.R., Minder, C.E., McMillan, I. and Burnside, E.B. (1977) Non-linear techniques for predicting 305-day lactation production for Holsteins and Jerseys. *Journal of Dairy Science* 60, 1636–1644.

Schaeffer, L.R., Jamrozik, J., Kistemaker, G.J. and Van Doormaal, B.J. (2000) Experience with a test-day model. *Journal of Dairy Science* 83, 1135–1144.

Stanton, T.L., Jones, L.R., Everett, R.W. and Kachman, S.D. (1992) Estimating milk, fat, and protein lactation curves with a test day model. *Journal of Dairy Science* 75, 1691–1700.

Wood, P.D.P. (1967) Algebraic model of the lactation curve in cattle. *Nature* 216, 164–165.

Wood, P.D.P. (1968) Factors affecting the shape of the lactation curve in cattle. *Animal Production* 11, 307–316.

15

Development and Commercialization of Software for Genetic Improvement Programmes: a Case Study

Brian Kinghorn,[1] Susan Meszaros[1] and John Wooliams[2]

[1]*Department of Animal Science , University of New England, Armidale, NSW 2351, Australia;* [2] *Roslin Institute (Edinburgh), Roslin EH25 9PS, UK*

Introduction

The development of computers within the last 20 years has been dramatic. Whilst computer scientists may have been making predictions of such changes, it would have been hard for livestock breeders to appreciate fully that, within a short space of time, populations of millions would have individual breeding values calculated using complex and demanding statistical techniques. Examination of pre-1990 research indicates the extent of computational development since. Many techniques that are common and accepted in research today were rare, e.g. simple stochastic simulation of populations and Bayesian analysis of complex data using Monte–Carlo Markov Chains. Even providing a standard error for a variance component was a challenge! This surge in computing power has prompted the development of concepts to remove important areas of ignorance in our understanding of genetic improvement schemes, and it is not surprising that this same surge has had a profound impact on their operational management.

As with many successful technologies, the application of computer software in the operational management of improvement programmes has had a positive feedback. The opportunities offered have been taken up and, in turn, suggested novel solutions and further opportunities. For example the development and application of software to calculate best linear unbiased predictors (BLUP) in commercial populations has not only been used as a means of calculating estimated breeding values (EBVs) to use in a selection index (its most immediate application), but has prompted improvements in the operational organization of evaluation (i.e. which animals are kept where) and genetic management (i.e. what is the best way of utilizing and conserving the genetic variation available).

Software has now become a central part of the process in improvement programmes in developed countries. Nevertheless, for the most part, the actual users of the software are breeding companies and/or organizations that are serving the top of the breeding pyramid. With continuing development in computer power (the PC in a dairy parlour considerably outperforms mainframe installations of 20 years ago) and the potential to move data and relay information vast distances instantaneously using the Internet, a challenge facing software developers in animal breeding is to make it more versatile and increasingly relevant to more of the breeding pyramid.

This chapter examines the commercialization of software relevant to the genetic improvement of livestock, in particular how commercialization is tackling the needs for versatility and broader relevance. Commercialization brings with it the requirement to protect the intellectual property (IP) of the authors and, hence, consideration of how best to deliver the services to satisfy these demands as well as good utility.

What Drives the Conception and Initiation of Software Development Projects?

Software development can be exciting. It is a way to take concepts and ideas and turn them into powerful, visual aids for teaching and for the development and successful implementation of breeding programmes. Software can help teach students the basics of genetic improvement theory. It can be a quick and powerful way to transform data to highly useful information, and it can help to make vital tactical decisions about selection and mating and underpin strategic planning. It is no wonder then that so many researchers and educators whose first focus is animal breeding turn their hand to the development of software for animal breeding.

The use of computers to do genetic analyses has led to the development of software for specific purposes. As a result, many students of genetics learn to write software programs with applications including population simulation, BLUP analyses, calculation of selection indices and other breeding programme 'design' issues. As the level of competence of geneticists in software development improves, and as the demand for more 'modern' software increases, we have seen the emergence of a number of quite sophisticated programs for a range of breeding issues.

Unplanned development by individuals and small groups

Aside from the enjoyment gained from writing code, software development occurs for several reasons. Firstly, a piece of software written for a specific research purpose may be found to have application in a 'commercial' manner, and is then adapted to be more useful. This is a natural progression from the development of software to carry out the latest research, sometimes contributing directly to animal breeding theory, through to the *application* of the research using the same software. It is natural because it is usually the researcher who best understands the problem in hand, and the best methods used to solve it. These methods are embedded in the software, and thus it is the most natural place to begin when using the fruits of the underlying science.

Secondly, individuals or small groups of geneticists and educators may use computers and software specifically written for teaching. These people develop the necessary software because they require it, and it often becomes quite sophisticated because of the need to deliver not only numerical results, but also graphical results in order to visualize the concepts involved. With the ability to deliver dynamic visual images in the classroom using 'beamers' connected to a computer, and to provide educational and reference material remotely via the Internet, there is an emergence of teaching software deployed using a number of methods. These are described in more detail below.

Software that has been developed primarily for educational purposes often finds commercial application. Small changes to the software enable it to be used by breeders and consultants to breeders for both learning and as breeding programme aids. Typically these are programs used to calculate optimal breeding structures, selection intensities, selection indices and the viewing of data such as pedigree data or lineage.

Many individuals or small groups develop software 'on the side', often without specific funds, and not working to set deadlines. They have the luxury of time and a less pressured environment and therefore are often more creative in the software they develop. The resulting software products tend to be smaller and more specialized than those developed by purpose-funded groups.

Planned development in purpose-funded groups

Some groups come together for the sole purpose of developing software for a specific project and/or client. This has generally been quite rare in the field of livestock improvement, but is occurring more often particularly with genetic evaluation software. The drive for such planned development by purpose-funded groups is an increasing need for software that is not only scientifically sound and useful, but also robust, reliable and easy to maintain and upgrade. There is now a small number of such groups around the world that develops, support and improve software.

Software such as BREEDPLAN (http://breedplan.une.edu.au/) is used to deliver BLUP EBVs to cattle breeders around the world and its deliverers depend upon its reliability, scientific soundness and integrity. Not only must the software reflect scientific principles and stand up to the rigour of scientific scrutiny, but it must also be stable, efficient and be able to be run effectively by commercial parties.

Although genetic evaluation software is among the most established, there is still demand for active development to provide better multi-trait software, that provides algorithms which better model both the genetic variation (e.g. threshold and quantitative trait locus (QTL) models) and environmental variation observed, and which cater for larger numbers of individuals in less computer time. This continuing demand suggests that these purpose-funded groups will not disappear overnight.

Some key factors in software development

Development of software for animal breeding is generally costly, requires key scientific and programming skills, must respond to a range of needs and often depends upon the desire of scientists to become 'software' developers. We will look at each of these factors briefly.

Whether or not software engineers are employed to write software alongside a scientist, writing software requires time. The investment of financial resources into professional programmers should be addressed with regard to the objective of the software development project and the requirements for more formal approaches to software development. Modern software development requires a wider range of skills than those held by all but a very few scientists. When the project is being funded by a third party, is being commercialized to third parties, or has significant cost or liability implications, it may be a wise use of resources to hire those whose specialty is software development. Their expertise lays not in the science, but in the design of the software itself, the choice of language and the development approach, and routes to effective deployment, including deployment via the Internet.

Professional software programmers or engineers will ensure the software has error checking, is maintainable and extendable, and runs efficiently on a number of platforms. However, if the software project is to be used in-house or it does not have criticality, then the use of software programmers may indeed be a luxury that cannot be afforded.

During training, animal breeders learn to write programs. Whether or not they learn to write programs well, and the range of skills they possess, is debatable. They are after all animal breeders! However, it is quite surprising the range of skills that do exist and how easily they can be obtained with a minimum of effort. Modern programming languages are highly visual in that they generally provide integrated development environments that not only allow programming of graphical user interfaces, but also take care of the mundane yet important task of revision control. As with the importance of using professional programmers, the requirement for a range of skills and abilities depends upon the software development objectives. Often a combination of a good scientist with some programming skills and a good software engineer with good software design skills is ideal for the development of good, commercial animal breeding software.

Of course, the objective of software development is to fulfil a need. Whether the need is driven from outside, that is from a contract or project, or from inside, a need to express one's creativity, problem or solution in code, there is generally a need – a problem to be solved. In the case of software destined for commercial use, the developers must be sure to address the need as expressed by the end users or clients, and not those they believe the needs to be. This may appear at first sight somewhat obvious; however, most software development projects fail because they do not address the real needs of users. Critical to this is the fact that technology, programming languages and the Internet among others, are changing rapidly. This means that what is possible to solve with software solutions also changes as technology changes. Not only do the problems themselves change, but also the way we expect to interact with the software changes. This factor is not to be taken too lightly, as it reflects the continually changing nature of software. Today, software is rarely finished! It is continually changing and adapting to the user's needs as they change and evolve.

The development of software for genetic improvement programmes requires not only the resources to make it happen: time, money and skills. It also requires animal breeders. It requires the knowledge and understanding of genetic principles that only highly trained animal breeders have. It requires insight into the problems, understanding of the principles and the ability to 'see' the application of the concepts. Successful software for genetic improvement programmes is a marriage of knowledge and skills, of desire and need, of developer and user.

IP and Commercialization Rights

What is IP? It is the smart ideas, more so than the hard work, that has made the software as good and useful as it is. Most animal breeding software is developed by scientists within universities and research organizations. These scientists like to be rich and famous, but usually not in that order. This means that their IP often ends up in the public domain. However, there is a notable trend away from this, often driven by the employers and funders of scientists, who seek assurances that valuable ideas be kept financially valuable to them.

IP is of little financial value if you do not have the right to commercialize it, or you do not have an appropriate royalty agreement with someone who does have the right to commercialize it. The group with appropriate commercialization rights calls the shots. It is usually in the interest of the IP holder(s) that there is a strong and business-orientated commercializer.

Paving the way for smooth deployment

Smooth and efficient deployment of good software can be severely hampered by IP and commercialization issues. This is especially so where there are several developers involved, and, more seriously, where there is more than one hosting entity (company, institution, funding agency, etc.).

An additional source of difficulty can be the access to data used as input by the software. A program may need access to databases on pedigree and/or estimates of breeding value, maybe in new formats or via novel portals, and there can be competing interests involved in the chain of data generation, analysis, storage and ownership.

There can be great value in paving the way for successful deployment, by negotiating a route through such political and commercial issues. Unfortunately, this can take considerable time – time that might otherwise be used for good research and development. It is incumbent on hosting institutions and funding agencies to provide for efficient and professional support in cases where such difficulties might arise.

Negotiating commercialization rights

In many cases, research programmes involving software development are funded by agencies that seek industry impact, rather than generating direct profit from use of software. However, the funded organization may plan a route to profits via delivery of software products and/or services. This situation should be recognized, and appropriate agreements on commercialization rights made at an early stage in the relationship between groups.

Such agreements should take account of the fact that method and software developed for a specific industry might be quite easily adapted to other industries, possibly competing industries.

Developer incentives

Scientists and programmers that develop successful software can be well motivated by the prospect of that success. However, where that success involves sizeable profit, as well as industry impact, appropriate rewards can help to escalate team excitement, cohesiveness and productivity. In some cases it may be practical to form a company around a group of successful software products, and give the developers an interest in that company.

Software Architecture and Patterns of Deployment

When planning the development of software solutions for genetic improvement programmes, it is critical to identify who the users of the software will be, and who will benefit from the results generated. In many cases, these will not be the same people or groups of people, for a number of reasons.

- The tasks required of computer programs in this area are becoming more diverse and complex. This makes it difficult to expect the beneficiaries (farmers and breeders) who are not expert computer users, or indeed expert in the technical areas involved, to be able to be in command of the full software implementation process.
- Many functions require connection to animal and related databases, which in some cases should be updated on a frequent basis. The Internet can be used to give a single user access to these databases, but the levels of competence and knowledge required are likely to be beyond most farmers or breeders.
- There can be added functionality in computer programs where they can be applied in coordination across a number of interrelated farming units. This might include online auction systems, group decision making in cooperative schemes and across-unit genetic evaluation.

On the other hand, the beneficiaries of the software results will tend to be the users of the software in cases such as:

- On-the-spot tactical decision tools, as might be used to identify the best backup male to use when a planned mating is not possible due to debility or death of the originally chosen male.
- Situations in which the beneficiary does not want any possible compromise of security of important information or data.

With good knowledge or understanding of who the users and beneficiaries of the proposed software system are, and with understanding of the desired functionality of the system, we can begin to consider ideal patterns of deployment.

Central analysis using a bureau service

The central bureau is the most common way of providing software-based information for genetic improvement throughout the breeding pyramid, and not just to the very top tier. Many genetic evaluation systems have been developed as bureau services, especially so in the dairy, beef and sheep industries where many independent farm enterprises are linked in across-group analyses. The reason for this is partly historical since: (i) older software was not designed for easy use, and it depended upon trained personnel to manage data input and interpret output, and (ii) only through central bureaux was it feasible to provide the necessary computing power and links to the required input data. However, neither of these limitations continues to apply and whether or not a central bureau is the most appropriate means to deliver a service is nowadays a serious design question.

Central bureau services have the great advantage of being able to generate large national databases, which give an excellent basis for the fair comparison of breeding stock raised in different farms, and even in different countries. These databases also provide an excellent resource for research.

Disadvantages of such schemes have involved unacceptable delays between delivery of information from the breeder and return of processed results for the breeder to use. Furthermore, the quality of the services provided to all beneficiaries will depend on the management ability of the single central bureau in setting both service and software standards. In some cases there may also be problems due to lack of credibility or 'ownership' of the results by the breeder, because s/he has not been involved directly in the process of analysis. This is why good education programmes can be essential for the success of central bureau services.

Distributed stand-alone executable software

Most existing software to help implement genetic improvement programmes is not suitable for use by breeders who do not have access to personnel trained in quantitative genetics. This category is thus largely filled with software targeted at quantitative geneticists, statisticians and appropriately trained consultants.

The range of function of these computer programs is very large, including:

- calculation of estimated breeding values;
- development of breeding objectives and relative economic weights;
- development of cross-breeding strategies;
- planning of breeding design;

- implementation of breeding design;
- estimation of statistical parameters;
- computer-aided learning programs.

Because the user operates the stand-alone program directly, or possibly with the aid of a consultant, or an extension officer, there is some scope for self-education on the technical basis of the issue(s) concerned, and for a greater identification with, and acceptance of, the results achieved. Of course, this type of software can require considerable user training, and ongoing support.

Internet deployment with central processing

There are increasing moves towards using the Internet to host and deploy software systems and services. Shown in Fig. 15.1 is a possible configuration. In this case the user communicates with a server to construct a data set, to drive the technical component of the software, and to gather and deliver results.

Internet deployment can give the benefits of both stand-alone software and bureau-served software, without many of the disadvantages, as suggested in Table 15.1. It can be set up to give the purveyors of the software or service immediate information on pattern and extent of use by individual users. This in turn can lead to relevant and timely user

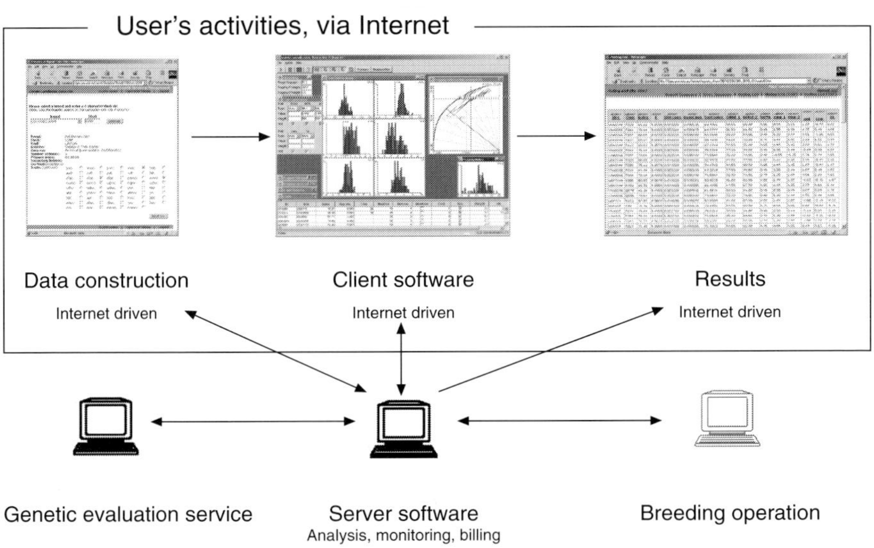

Fig. 15.1. A possible configuration for Internet deployment with central processing.

Table 15.1. A contrast of different methods for providing software and services.

	Central bureau service	Stand-alone software	Internet deployed software/service
User independence	Very low	Very high	High
Speed of achieving results	Low	High	High
Typical level of user support	Low	Medium	High
Ease of upgrading	High	Low	High (can be automatic)
Ease of coordinated analysis across farming units	High	Low	High
Ease of connection to relevant information and support	Medium	Medium	High
Ease of exploiting intermediate consultants	High	Medium	High
Key examples	Breedplan	PEST	TGRM
	Lambplan	Pigblup	QTL Express

support, and even automatic billing based on objective criteria. It is also possible to disseminate software upgrades with little or zero user intervention (see, for example, http://java.sun.com/products/javawebstart/).

There is also the potential to connect peer users in breeding games (B. Golden, personal communication), and in software systems that support buying and selling of genetic material – computer aided seedstock trading, CAST (Kinghorn, 1989).

Internet deployment with local processing

This category is currently represented by the increasing number of Java Applets that are becoming available on Web sites. One example is the Java version of the learning aid 'Sel' (http://www-personal.une.edu.au/~bkinghor/sel.htm), which runs a simple visually orientated stochastic simulation of a breeding programme. Coupled with good on-line reference material, such as technical notes, help files, exercises and quizzes, this type of software is set to play a major role in educating animal breeding students and practitioners.

Conclusions

It is clear that there is a growing development of software for managing genetic improvement programmes and this has been stimulated by farm-based computing power and modern software tools. For the most part, the software that is being developed is stand-alone software, often supported by research programmes receiving public funds as a means of providing a clear route for technology transfer, although there are notable exceptions.

The scope of the available software encompasses research that often: (i) utilizes very recent genetic concepts and models to provide solutions (e.g. combining selection with the management of genetic variation, QTL detection), (ii) requires the development of computing algorithms to deliver the solutions (e.g. BLUP, management of genetic variation), (iii) has required the bringing together of different disciplines (e.g. genetics with computing science, economics and statistics). This development, whilst prompted by demand and need, has depended upon research environments to provide solutions.

The development and commercialization of software packages poses a serious dilemma for research organizations. Whilst it provides an increasingly effective route for technology transfer, it brings with it a burden of customer support. In the first instance this will often mean diverting some of the research organization's resources for this purpose, and away from research activity. This has consequences for the research organizations, since a commercial culture is needed as distinct from a research culture, and research will be inhibited from developing solutions to the many research challenges that remain. Partnerships between levy boards and research organizations provide a possible solution to this dilemma (e.g. BREEDS, TGRM).

There is a temptation to view a specific piece of software, designed to carry out a specific task, as embodying *the* solution to the problem, and the problem is therefore a problem no longer. This is a dangerous assumption: software, even assuming it is doing something sensible, will frequently implement only one of a number of current solutions to a problem, each with their limitations. However, the perception of an all-embracing solution has the potential to slow development of the science, unless the development is fuelled by competition and investment from the organizations undertaking the commercialization. Therefore there is a question over the impact of commercialization of genetic software on the development of the science of livestock breeding. This would undoubtedly be overcome if, as in the pharmaceutical industry, the return from commercialization were sufficiently great.

As yet, it remains uncertain as to whether genetic software in livestock breeding can be provided as a continuing and profitable service. Sustainability as a commercial service will usually require extension down the breeding pyramid beyond the top tier. As a comparison, software to support economic management on farms (a key issue to all farmers!) *has* been able to develop as a service, but this has often been achieved by integrating the service with other agricultural service providers such as recording organizations and breed societies. It is likely that the future of genetic software will follow the same route and such developments are already apparent (e.g. LAMBPLAN).

The need for integration has wider perspectives: (i) in many countries the environmental demands made upon farmers will dictate the close integration of genetics with other aspects of management; and (ii) reducing subsidies are increasing unit size and squeezing labour, making computerized information gathering and management more attractive. Both these trends suggest that integrated management software will play an increasing role in successful farm management and this offers an opportunity for genetic software.

However, it has often been the case that genetics on the farm has often been left to the 'consultant' or the 'agent for the semen company' with the result that genetics capability is often limited. Therefore an additional challenge for unlocking the potential of genetic software throughout the pyramid, and in doing so realizing the full benefits of genetic improvement in our livestock populations, is to ensure that education in genetics is strong and accommodates the high potential offered by the software revolution.

References

Groeneveld, E., Kovac, M. and Wang, T. (1990) PEST, a general purpose BLUP package for multivariate prediction and estimation. In: *4th World Congress on Genetics Applied to Livestock Production*, Edinburgh, number XIII, pp. 488–491.

Kinghorn, B.P. (1989) Exploiting the total genetic resource in beef cattle. *Proceedings of Beef Breeding Workshop*, South Australian Research and Development Institute, Adelaide, 23–26 October 1989, Paper 11.

Newman, S., Stewart, T.S., Goddard, M.E. and Gregory, M. (1997) HotCross – a decision support aid for crossbreeding of beef cattle in tropical and subtropical environments. *Proceedings of the 12th Association for the Advancement of Animal Breeding and Genetics Conference*, Dubbo, April. AAABG, University of New England, Armidale, New South Wales, pp. 400–403.

Appendix 15.1. Some Examples of Software for Genetic Improvement Programmes

This appendix lists some example software, chosen to cover a range of applications and delivery types. For more complete listings, see the following:

http://www.math.usu.edu/~vukasino/ANSCI/soft_index.htm
http://nitro.biosci.arizona.edu/zbook/book.html#Links

The Animal Breeders Toolkit (ABTK) is a collection of software components that computer literate quantitative geneticists can use to construct programs with a variety of functions, including genetic evaluation and parameter estimation. The ABTK is available from ftp://cgel.Agsci.Colostate.EDU/pub/abtk. See also http://nce.ads.uga.edu/~ignacy/programs.html for a similar example.

ASREML is a mixed model analysis package developed by New South Wales Department of Agriculture and Rothampstead Experimental Station (http://www.chiswick.anprod.csiro.au/lists/asreml/). Delivery is as stand-alone software, with source code written in FORTRAN. The target user in genetic improvement programmes is the top tier of the breeding pyramid and research organizations. Whilst there are a number of mixed model packages available the particular value of ASREML is the complexity of the models that can be fitted.

BREEDPLAN (http://breedplan.une.edu.au/) is a bureau service for calculation of BLUP EBVs in beef cattle. In many cases the system is integrated with breed societies' pedigree system, giving whole-breed analyses.

BREEDS (http://www.ed.sac.ac.uk/beef_BREEDS) is a software package being developed to evaluate alternative replacement breeding policies for suckler beef herds in the UK developed by Scottish Agricultural College, MLC and Abacus Biotechnology Ltd (NZ). It helps beef producers choose the best replacement policy, and, choose the most appropriate breeds or crosses for their production system, combining both genetic information on breed and cross-breed performance and farm-specific economic information. Delivery is as stand-alone software written in Visual Basic. The target user extends throughout the breeding pyramid, particularly beyond the top tier (which might be more concerned with the improvement of the specific breeds).

GENCONT (e.g. http://www.ipg.nl/pdf/fokkerij/poster%20EHJM%20EAAP2000.pdf) was written by ID-DLO to implement the work of Meuwissen and co-workers on utilizing genetic contributions as an aid to tactical selection and mating decisions. Delivery is as stand-alone software with source code written in Fortran. Its target users are breeding organizations, most likely in the top tier of the breeding pyramid, with some technical capability in-house, as it requires to be provided with the necessary input information summarizing breeding values and pedigrees.

HOTCROSS (Newman *et al.*, 1997) is a decision-aid tool for crossbreeding strategies in beef cattle. It predicts the performance of genotypes, adjusted for four environmental stresses: heat, ticks, worms, and nutrition, and the user can build a crossing strategy in a visual manner. Code from HOTCROSS and GENUP (http://metz.une.edu.au/~bkinghor/genup.htm) has been used to write CROSSSHEEP (http://www.woolpro.co.nz/RD-Genetics.pdf) with functionality similar to HOTCROSS. HOTCROSS is available as stand-alone software with source code written in Visual Basic. Target users are beef producers seeking to plan crossing systems that are both efficient and relevant to their conditions. Data and parameterization can limit target users to specific regions.

LAMBPLAN (http://lambplan.une.edu.au/) is a recording system and bureau service for calculation of BLUP EBVs in sheep. The system also delivers support on issues such as breeding programme design.

PEST (for Prediction and ESTimation) was written by Groeneveld *et al.* (1990) (see also http://www.tzv.fal.de/~eg/). It is a stand-alone software package for multivariate prediction and estimation – principally for estimating the breeding values of animals. It is a mature product, written in Fortran 77, with the last noted revision being 3 June 1993 (as noted in the manual current in November 2001). Target users are computer literate geneticists. PEST is still widely used as the heart of genetic evaluation systems, most notably in the pig industry.

PIGBLUP (http://agbu.une.edu.au/PIGBLUP/) is a stand-alone genetic evaluation program for pigs. It includes calculation of economic indices, genetic trends, genetic auditing, mate selection to avoid progeny inbreeding and screening of data for integrity. PIGBLUP is available through the Agricultural Business Research Institute (http://abri.une.edu.au/pigblup.htm). Target users are pig breeding nucleus operations.

QTL EXPRESS (http://qtl.cap.ed.ac.uk) is a Java web-based package for QTL analysis using regression interval mapping techniques developed by the University of Edinburgh and Roslin Institute. It currently supports analyses of F2 data from crosses between inbred or outbred lines, and data from outbred half-sib populations. Both input and output are via graphical interfaces; with input either linked to a user database or simply via text files, and output provides information on estimates, analysis and diagnostics. The target users are research organizations and breeding organizations in the top tier of the pyramid.

SELACTION (http://www.zod.wau.nl/vf/selaction/) is stand-alone visual software developed by the University of Wageningen and the Roslin Institute for the purpose of strategic design of genetic improvement programmes. The software deals with predicting gains and rates of inbreeding in selected populations. It is available as stand-alone software. Its intended users are breeding organizations in the top tier of the breeding pyramid and those engaged in experimental programmes.

SELIND was initially developed by E.P. Cunningham in the 1970s, to calculated selection index weights for multi-trait selection. It is written in Fortran and is targeted at computer literate quantitative geneticists.

TGRM, standing for Total Genetic Resource Management (http://tgrm.une.edu.au/), is a software-based service that is intended to provide tactical help in selection and mating decisions based on EBVs, pedigree information, information on specific genes, and prevailing logistical and cost constraints. Particular efforts have been made to provide for highly graphical and interactive interfaces. The package is written in JAVA and FORTRAN and is deployed via the Internet. The target users include nucleus and multiplier levels of all animal industries.

16

Final Thoughts

Max Rothschild and Scott Newman

Iowa State University and PIC USA

The use of artificial intelligence to determine breeding plans, sequenced genomes, transgenic livestock and cloned animals will probably become the norm during the 21st century. These discoveries and their uses represent the intellectual property (IP) of the individuals, scientific teams and research enterprises that develop the IP and commercialize it. Bruce Lehman, former US Commissioner of Patents and Trademarks, has said, 'The only wealth there is in the world is the wealth that comes from the human mind'. As has been discussed in the previous chapters, animal breeders have begun to patent their IP. Patenting of IP has raised economic, legal and ethical concerns that might affect the support of public education and research. These issues are far from settled; certainly public perceptions of science and the willingness to fund it will play a larger role in the future. Changes are under way at most universities, but the rush to patent all IP has been tempered by the difficulty in obtaining sufficient royalties to support such activities. However, the pursuit for royalties has allowed patents to aid technology transfer instead of blocking the spread of knowledge or stifling innovation.

Patent law is not fixed nor is it perfect. The extent of claims granted varies with time. For example, in the early 1990s, gene marker patents were extensive and claims included linked markers and extension to species other than those tested. By early 2000, gene marker patents in the USA were limited generally to the individual mutations, primers and sequences, and species of interest. It is certainly possible that the test day model patent discussed in Chapter 14 is likely not to have been granted today due to changes in patent law related in part to patenting of algorithms (see Chapter 9). Certainly the structure of patent approval adds to the vagaries in claims as patent examiners and their reviews of similar patents vary. Ultimately, the court often has the final say on the extent of claims and the validity of some inventions.

The 'landscape' of agriculture has changed with increased vertical integration of production and there is more control from the farm to the consumer's table. Inventions can move more quickly into the market place in an integrated system. However, certain production sectors, like the family farm, may be disadvantaged due to the limited availability of technology transfer and economies of scale related to adoption of expensive technology. The public is concerned about the safety of food products resulting from genetic manipulation. They must be assured that patenting by itself does not create unhealthy food. Prohibiting animal-related patents will not, however, rescue the family farm. Instead, consideration of ways to improve technology transfer in all sectors of agriculture is likely to be more effective.

Patent applications must not be frivolous in order that the real costs of and profits from patenting remain reasonable. There is rising evidence that the number of patents in the USA has begun to level off and even show some decline. Undoubtedly a reasoned look combined with the realities of the difficulty and expense in obtaining patents has taken hold. Moreover, there is some evidence to date that some patented technologies are being licensed at costs that the public is not willing to bear. A recent example of this appeared in *Nature* (7 February 2002) pertaining to the diagnostic test for haemochromatosis, in which several medical laboratories stopped performing the medical test, in part due to the large up-front royalties charged. Owners of patents must take care to charge reasonable royalties in order that they can recover the investment but must not let the cost put the technology outside the reach of the public. Bundling of technologies and royalty payments are likely to aid in keeping technologies affordable. Reinvestment of a share of the profits from patenting must be used to promote future discovery. This is certainly true in the public research universities but must also take place in private corporations. Such reinvestment insures a healthy economy and promotion of science.

The next decade offers the promise of great advances related to our understanding of the genetics of our domestic species and methods to alter and manipulate their physiology to advance humankind. Scientists must work to see that the IP that is produced has usefulness and does not harm animals or humans. Finally, we must ensure that the quest to make new discoveries and patent the resulting IP is in the public good.

Index